BOB MILLER'S BASIC MATH AND PREALGEBRA

BASIC MATH
AND PREALGEBRA

BOB MILLER'S BASIC MATH AND PREALGEBRA

BASIC MATH AND PREALGEBRA

Robert Miller

Mathematics Department
City College of New York

McGraw-Hill

New York   Chicago   San Francisco
Lisbon   London   Madrid   Mexico City   Milan
New Delhi   San Juan   Seoul   Singapore
Sydney   Toronto

BOB MILLER'S BASIC MATH AND PREALGEBRA FOR THE CLUELESS

2 3 4 5 6 7 8 9 10 11 12 13 14 15 16 17 18 19 20 DOC DOC 0 9 8 7 6 5 4 3 2

ISBN 0-07-139016-2

Sponsoring Editor: Barbara Gilson
Production Supervisor: Clara Stanley
Editing Supervisor: Maureen Walker
Compositor: North Market Street Graphics
Photo: Eric Miller

Library of Congress Cataloging-in-Publication Data applied for.

McGraw-Hill

*A Division of The **McGraw·Hill** Companies*

To my wonderful wife Marlene.
I dedicate this book and everything else I ever do to you.
I love you very, very much.

TO THE STUDENT

This book is written for you: not for your teacher, not for your neighbor, not for anyone but you.

This book is written for those who want to get a jump on algebra and for those returning to school, perhaps after a long time.

The topics include introductions to algebra, geometry, and trig, and a review of fractions, decimals, and percentages, and several other topics.

In order to get maximum benefit from this book, you must practice. Do many exercises until you are very good with each of the skills.

As much as I hate to admit it, I am not perfect. If you find anything that is unclear or should be added to the book, please write to me c/o Editorial Director, McGraw-Hill Schaum Division, Two Penn Plaza, New York, NY 10121. Please enclose a self-addressed, stamped envelope. Please be patient. I will answer.

After this book, there are the basic books such as *Algebra for the Clueless* and *Geometry for the Clueless*. More advanced books are *Precalc with Trig for the Clueless* and *Calc I, II, and III for the Clueless*. For those taking the SAT, my *SAT Math for the Clueless* will do just fine.

Now Enjoy this book and learn!!

ACKNOWLEDGMENTS

I would like to thank my editor, Barbara Gilson, for redesigning my books and expanding this series to eight. Without her, this series would not be the success it is today.

I also thank Mr. Daryl Davis. He and I have a shared interest in educating America mathematically so that all of our children will be able to think better. This will enable them to succeed at any endeavor they attempt. I hope my appearance with him on his radio show "Our World," WLNA 1420, in Peekskill, N.Y., is the first of many endeavors together.

I would like to thank people who have helped me in the past: first, my wonderful family who are listed in the biography; next, my parents Lee and Cele and my wife's parents Edith and Siebeth Egna; then my brother Jerry; and John Aliano, David Beckwith, John Carleo, Jennifer Chong, Pat Koch, Deborah Aaronson, Libby Alam, Michele Bracci, Mary Loebig Giles, Martin Levine of Market Source, Sharon Nelson, Bernice Rothstein, Bill Summers, Sy Solomon, Hazel Spencer, Efua Tonge, Maureen Walker, and Dr. Robert Urbanski of Middlesex County Community College.

As usual the last thanks go to three terrific people: a great friend Gary Pitkofsky, another terrific friend and fellow lecturer David Schwinger, and my sharer of dreams, my cousin Keith Robin Ellis.

CONTENTS

CONGRATULATIONS

Congratulations!!!! You are starting on a great adventure. The math you will start to learn is the key to many future jobs, jobs that do not even exist today. More important, even if you never use math in your future life, the thought processes you learn here will help you in everything you do.

I believe your generation is the smartest and best generation our country has ever produced, and getting better each year!!!! Every book I have written tries to teach serious math in a way that will allow you to learn math without being afraid.

In math there are very few vocabulary words compared to English. However, many occur at the beginning. Make sure you learn *and* understand each and every word. Let's start.

BASIC MATH
AND PREALGEBRA

CHAPTER 1

INTRODUCTORY TERMS

At the beginning, we will deal with two sets of numbers. The first is the set of natural numbers, abbreviated by *nn,* which are the numbers 1, 2, 3, 4, . . . and the whole numbers 0, 1, 2, 3, 4, ... The three dots at the end means the set is INFINITE, that it goes on forever.

We will talk about equality statements, such as 2 + 5 = 7, 9 − 6 = 3 and *a* − *b* = *c.* We will write 3 + 4 ≠ 10, which says 3 plus 4 does not equal 10. −4, $\sqrt{7}$, π, and so on are not natural numbers and not whole numbers.

(3) · (4) = 12. 3 and 4 are FACTORS of 12 (so are 1, 2, 6, and 12).

A PRIME natural number is a natural number with two distinct natural number factors, itself and 1. 7 is a prime because only (1) × (7) = 7. 1 is not a prime. 9 is not a prime since 1 × 9 = 9 and 3 × 3 = 9. 9 is called a COMPOSITE. The first 8 prime factors are 2, 3, 5, 7, 11, 13, 17, and 19.

The EVEN natural numbers is the set 2, 4, 6, 8, ...

The ODD natural numbers is the set 1, 3, 5, 7, ...

We would like to graph numbers. We will do it on a LINE GRAPH or NUMBER LINE. Let's give some examples.

EXAMPLE 1

Graph the first four odd natural numbers.

First, draw a straight line with a ruler.

Next, divide the line into convenient lengths.

Next, label 0, called the ORIGIN, if practical.

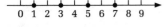

Finally, place the dots on the appropriate places on the number line.

EXAMPLE 2

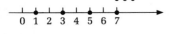

Graph all the odd natural numbers.

The three dots above mean the set is infinite.

EXAMPLE 3

Graph all the natural numbers between 60 and 68.
    The word "between" does NOT, NOT, NOT include the end numbers.
    In this problem, it is not convenient to label the origin.

EXAMPLE 4

Graph all the primes between 40 and 50.

EXAMPLE 5

Graph all multiples of 10 between 30 and 110 inclusive.

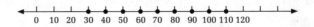

Inclusive means both ends are part of the answer.

Natural number multiples of 10: take the natural numbers and multiply each by 10.

Because all of these numbers are multiples of 10, we divide the number line into 10s.

A VARIABLE is a symbol that changes. In the beginning, most letters will stand for variables.

A CONSTANT is a number that does not change. Examples are 9876, π, 4/9, . . . , are all symbols that don't change.

We also need words for addition, subtraction, multiplication, and division.

Here are some of the most common:

Addition: sum (the answer in addition), more, more than, increase, increased by, plus.

Subtraction: difference (the answer in subtraction), take away, from, decrease, decreased by, diminish, diminished by, less, less than.

Multiplication: product (the answer in multiplication), double (multiply by 2), triple (multiply by 3), times.

Division: quotient (the answer in division), divided by.

Let's do some examples to learn the words better.

EXAMPLE 6—

The sum of p and 2. Answer: $p + 2$ or $2 + p$.

The order does not matter because of the COMMUTATIVE LAW of ADDITION which says the order in which you add does not matter. $c + d = d + c$. $84 + 23 = 23 + 84$.

The wording of subtraction causes the most problems. Let's see.

EXAMPLE 7—

A. The difference between x and y	$x - y$
B. x decreased by y	$x - y$
C. x diminished by y	$x - y$
D. x take away y	$x - y$
E. x minus y	$x - y$
F. x less y	$x - y$
G. x less than y	$y - x$
H. x from y	$y - x$

Very important. Notice "less" does NOT reverse whereas "less than" reverses. 6 less 2 is 4 whereas 6 less than 2 is $2 - 6 = -4$, as we will see later. As you read each one, listen to the difference!!

Also notice division is NOT commutative since $7/3 \neq 3/7$.

EXAMPLE 8—

The product of 6 and 4. 6(4) or (6)(4) or (6)4 or (4)6 or 4(6) or (4)(6).

Very important again. The word "and" does NOT mean addition. Also see that multiplication is commutative. $ab = ba$. $(7)(6) = (6)(7)$.

EXAMPLE 9—

A. Write s times r; B. Write m times 6. Answers: A. rs; B. $6m$.

A. Although either order is correct, we usually write products alphabetically.

B. Although either order is again correct, we always write the number first.

EXAMPLE 10—

Write 2 divided by r. Answer: $\dfrac{2}{r}$.

For algebraic purposes, it is almost always better to write division as a fraction.

EXAMPLE 11—

Write the difference between b and c divided by m.

Answer: $\dfrac{b-c}{m}$.

EXAMPLE 12—

Write: two less the sum of h, p, and m. Answer: $2 - (h + p + m)$.

This symbol, ( ), are parentheses, the plural of parenthesis.

[ ] are brackets. { } are braces.

There are shorter ways to write the product of identical factors. We will use EXPONENTS or POWERS.

y^2 means $(y)(y)$ or yy and is read "y squared" or "y to the second power." The 2 is the exponent or the power.

8^3 means $8(8)(8)$ and is read "8 cubed" or "8 to the third power."

x^4 means $xxxx$ and is read "x to the fourth power."

x^n $(x)(x)(x) \ldots (x)$ [x times (n factors)] and is read "x to the nth power."

$x = x^1$, x to the first power.

I'll bet you weren't expecting a reading lesson. There are always new words at the beginning of any new subject. There are not too many later, but there are still some more now. Let's look at them.

$5x^2$ means $5xx$ and is read "5, x squared."

$7x^2y^3$ is $7xxyyy$, and is read "7, x squared, y cubed."

$(5x)^3$ is $(5x)(5x)(5x)$ and is read "the quantity $5x$, cubed." It also equals $125x^3$.

EXAMPLE 13—

Write in exponential form. (Write with exponents; do NOT do the arithmetic!)

A. $(3)(3)(3)yyyyy$; B. $aaabcc$;
C. $(x + 6)(x + 6)(x + 6)(x - 3)(x - 3)$.

Answers: A. $3^3 \, y^5$; B. a^3bc^2; C. $(x + 6)^3(x - 3)^2$.

EXAMPLE 14—

Write in completely factored form with no exponents:
A. $84 \, a^4bc^3$; B. $30(x + 6)^3$.

Answers: A. $(2)(2)(3)(7)aaaabccc$;
B. $(2)(3)(5)(x + 6)(x + 6)(x + 6)$.

ORDER OF OPERATIONS, NUMERICAL EVALUATIONS

Suppose we have $2 + 3(4)$. This could mean $5(4) = 20$ orrrr $2 + 12 = 14$. Which one? In math, this is definitely a no no! An expression can have one meaning and one meaning only. The ORDER OF OPERATIONS will tell us what to do first.

1. Do any operations inside parenthesis or on the tops and bottoms of fractions.

2. Evaluate numbers with exponents.

3. Multiplication and division, left to right, as they occur.

4. Addition and subtraction, left to right.

PEMDAS

EXAMPLE 1—

Our first example $2 + 3(4) = 2 + 12 = 14$ since multiplication comes before addition.

EXAMPLE 2—

$5^2 - 3(5 - 3) + 2^3$ inside parenthesis first

$= 5^2 - 3(2) + 2^3$ exponents

$= 25 - 3(2) + 8$ multiplication, then adding and subtracting

$= 25 - 6 + 8 = 27$

EXAMPLE 3—

$24 \div 8 \times 2$ Multiplication and division, left to right, as they occur. Division is first. $3 \times 2 = 6$.

EXAMPLE 4—

$$\frac{4^3 + 6^2}{12 - 2} + \frac{8(4)}{18 - 2} = \frac{64 + 36}{12 - 2} - \frac{8(4)}{18 - 2} = \frac{100}{10} + \frac{32}{16}$$

$$= 10 + 2 = 12.$$

Sometimes we have a step before step 1. Sometimes we are given an ALGEBRAIC EXPRESSION, a collection of factors and mathematical operations. We are given numbers for each variable and asked to EVALUATE, find the numerical value of the expression. The steps are . . .

1. Substitute in parenthesis, the value of each letter.

2. Do inside of parentheses and the tops and bottoms of fractions.

3. Do each exponent.

4. Do multiplication and division, left to right, as they occur.

5. Last, do all adding and subtracting.

EXAMPLE 5—

If $x = 3$ and $y = 2$, find the value of: A. $y(x + 4) - 1$; B. $5xy - 7y$; C. $x^3y - xy^2$; D. $\dfrac{x^4 - 1}{x^2 - y^2}$.

A. $y(x + 4) - 1 = (2)\,[(3) + 4] - 1 = (2)(7) - 1 = 14 - 1 = 13$.

B. $5xy - 7y = 5(3)(2) - 7(2) = 30 - 14 = 16$.

C. $x^3y - xy^2 = (3)^3(2) - (3)(2)^2 = (27)(2) - 3(4) = 54 - 12 = 42$.

D. $\dfrac{x^4 - 1}{x^2 - y^2} = \dfrac{(3)^4 - 1}{(3)^2 - (2)^2} = \dfrac{81 - 1}{9 - 4} = \dfrac{80}{5} = 16$.

SOME DEFINITIONS, ADDING AND SUBTRACTING

We need a few more definitions.

TERM: Any single collection of algebraic factors, which is separated from the next term by a plus or minus sign. Four examples of terms are $4x^3y^{27}$, x, $-5tu$, and 9.

A POLYNOMIAL is one or more terms where all the exponents of the variables are natural numbers.

MONOMIALS: single-term polynomials: $4x^2y$, $3x$, $-9t^6u^7v$.

BINOMIALS: two-term polynomials: $3x^2 + 4x$, $x - y$, $7z - 9$, $-3x + 2$.

TRINOMIALS: three-term polynomials: $-3x^2 + 4x - 5$, $x + y - z$.

COEFFICIENT: Any collection of factors in a term is the coefficient of the remaining factors.

If we have $5xy$, 5 is the coefficient of xy, x is the coefficient of $5y$, y is the coefficient of $5x$, $5x$ is the coefficient of y, $5y$ is the coefficient of x, and xy is the coefficient of 5. Whew!!!

Generally when we say the word coefficient, we mean NUMERICAL COEFFICIENT. That is what we will use throughout the book unless we say something else. So the coefficient of $5xy$ is 5. Also the coefficient of $-7x$ is -7. The sign is included.

The DEGREE of a polynomial is the highest exponent of any one term.

EXAMPLE I—

What is the degree of $-23x^7 + 4x^9 - 222$? The degree is 9.

EXAMPLE 2—

What is the degree of $x^6 + y^7 + x^4y^5$?

The degree of the x term is 6; the y term is 7; the xy degree is 9 $(= 4 + 5)$.

The degree of the polynomial is 9.

We will need only the first example almost all the time.

EXAMPLE 3—

Tell me about $5x^7 - 3x^2 + 5x$.

1. It is a polynomial since all the exponents are natural numbers.

2. It is a trinomial since it is three terms.

3. $5x^7$ has a coefficient of 5, a BASE of x, and an exponent (power) of 7.

4. $-3x^2$ has a coefficient of -3, a base of x, and an exponent of 2.

5. $5x$ has a coefficient of 5, a base of x, and an exponent of 1.

6. Finally, the degree is 7, the highest exponent of any one term.

EXAMPLE 4—

Tell me about $-x$.

It is a monomial. The coefficient is -1. The base is x. The exponent is 1. The degree is 1.

$-x$ really means $-1x^1$. The ones are not usually written. If it helps you in the beginning, write them in.

In order to add or subtract, we must have like terms.

LIKE TERMS are terms with the exact letter combination AND the same letters must have identical exponents.

$y = y$ is called the reflexive law. An algebraic expression always = itself.

We know $y = y$ and $abc = abc$. Each pair are like terms.

a and a^2 are not like terms since the exponents are different.

x and xy are not like terms.

$2x^2y$ and $2xy^2$ are not like terms since $2x^2y = 2xxy$ and $2xy^2 = 2xyy$.

As pictured, $3y + 4y = 7y$.

	$3y$		$4y$	
		$7y$		

Also $7x^4 - 5x^4 = 2x^4$.

To add or subtract like terms, add or subtract their coefficients; leave the exponents unchanged.

Unlike terms cannot be combined.

EXAMPLE 5A—

Simplify $5a + 3b + 2a + 7b$. Answer: $7a + 10b$. Why? See Example 5b.

EXAMPLE 5B—

Simplify 5 apples + 3 bananas + 2 apples + 7 bananas.
Answer: 7 apples + 10 bananas. Well you might say 17
pieces of fruit. See Example 5c.

EXAMPLE 5C—

Simplify 5 apples + 3 bats + 2 apples + 7 bats. Answer:
7 apples + 10 bats.

Only like terms can be added (or subtracted). Unlike
terms cannot be combined.

EXAMPLE 6—

Simplify $4x^2 + 5x + 6 + 7x^2 - x - 2$.
Answer: $11x^2 + 4x + 4$.

Expressions in one variable are usually written highest
exponent to lowest.

EXAMPLE 7—

Simplify $4a + 9b - 2a - 6b$. Answer: $2a + 3b$. Terms are
usually written alphabetically.

EXAMPLE 8—

Simplify $3w + 5x + 7y - w - 5x + 2y$. Answer: $2w + 9y$.
$5x - 5x = 0$ and is not written.

Commutative law of addition: $a + b = b + a$; $4x + 5x =$
$5x + 4x$.

Associative law of addition: $a + (b + c) = (a + b) + c$;
$(3 + 4) + 5 = 3 + (4 + 5)$.

We will deal a lot more with minus signs in the next
chapter.

After you are well into this book, you may think
these first pages were very easy. But some of you may
be having trouble because the subject is so very, very
new. Don't worry. Read the problems over. Solve
them yourself. Practice in your textbook. Everything
will be fine!

PRODUCTS, QUOTIENTS, AND THE DISTRIBUTIVE LAW

Suppose we want to multiply a^4 by a^3:

$$(a^4)(a^3) = (aaaa)(aaa) = a^7$$

Products

LAW 1   If the bases are the same, add the exponents. In symbols, $a^m a^n = a^{m+n}$.

1. The base stays the same.

2. Terms with different bases and different exponents cannot be combined or simplified.

3. Coefficients are multiplied.

EXAMPLE 1—

$a^6 a^4$. Answer: a^{10}.

EXAMPLE 2—

$b^7 b^4 b$. Answer: b^{12}. Remember: $b = b^1$.

EXAMPLE 3—

$(2a^3 b^4)(5b^6 a^8)$. Answer: $10a^{11}b^{10}$.

1. Coefficients are multiplied $(2)(5) = 10$.

2. In multiplying with the same bases, the exponents are added: $a^3 a^8 = a^{11}$, $b^4 b^6 = b^{10}$.

3. Unlike bases with unlike exponents cannot be simplified.

4. Letters are written alphabetically to look pretty!!!!

EXAMPLE 4—

$3^{10}3^{23}$. Answer: 3^{33}. The base stays the same.

Order is alphabetical although the order doesn't matter because of the commutative law of multiplication and the associative law of multiplication.

Commutative law of multiplication: $bc = cb$.

$(3)(7) = (7)(3)$ $(4a^2)(7a^4) = (7a^4)(4a^2) = 28a^6$

Associative law of multiplication: $(xy)z = x(yz)$.

$(2 \cdot 5)3 = 2(5 \cdot 3)$ $(3a \cdot 4b)(5d) = 3a(4b \cdot 5d) = 60abd$

EXAMPLE 5—

Simplify $4b^2(6b^5) + (3b)(4b^3) - b(5b^6)$

SOLUTION—

$24b^7 + 12b^4 - 5b^7$ Order of operations; multiplication first.

$= 19b^7 - 12b^4$ Only like terms can be combined; unlike terms can't.

Suppose we have $(a^4)^3$. $(a^4)^3 = a^4a^4a^4 = a^{12}$.

A power to a power? We multiply exponents. In symbols. . . .

LAW 3 $(a^m)^n = a^{mn}$. Don't worry, we'll get back to law 2.

EXAMPLE 6—

$(10x^6)^3(3x)^4$.
$(10x^6)^3(3x)^4 = (10^1x^6)^3(3^1x^1)^4 = 10^3x^{18}3^4x^4 = 81,000x^{22}$.

LAW 4 $(ab)^n = a^nb^n$.

EXAMPLE 7—

$(a^3b^4)^5 = a^{15}b^{20}$.

QUOTIENTS: $\dfrac{a}{b} = c$ if $a = bc;$ $\dfrac{12}{3} = 4$

because $12 = 3(4)$.

THEOREM (A PROVEN LAW)   Division by 0 is not allowed.

   Whenever I teach a course like this, I demonstrate and show everything, but I prove very little. However this is too important not to prove. Zero causes the most amount of trouble of any number. Zero was a great discovery, in India, in the 600s. Remember Roman numerals had no zero! We must know why 6/0 has no meaning, 0/0 can't be defined, and 0/7 = 0.

Proof   Suppose we have $\dfrac{a}{0}$, where $a \neq 0$.

If $\dfrac{a}{0} = c$, then $a = 0(c)$. But $o(c) = 0$. But this means $a = 0$. But we assumed $a \neq 0$. So assuming $a/0 = c$ could not be true. Therefore expressions like 4/0 and 9/0 have no meaning.

If $\dfrac{0}{0} = c$, then $0 = 0(c)$. But c could be anything. This is called indeterminate.

But 0/7 = 0 since $0 = 7 \times 0$.

By the same reasoning $\dfrac{x^5}{x^2} = x^3$, since $x^5 = x^2x^3$.

Looking at it another way $\dfrac{x^5}{x^2} = \dfrac{\overset{1}{\cancel{x}}\,\overset{1}{\cancel{x}}\,x\,x\,x}{\underset{1}{\cancel{x}}\,\underset{1}{\cancel{x}}} = x^3$.

Also $\dfrac{y^3}{y^7} = \dfrac{\overset{1}{\cancel{y}}\,\overset{1}{\cancel{y}}\,\overset{1}{\cancel{y}}}{\underset{1}{\cancel{y}}\,\underset{1}{\cancel{y}}\,\underset{1}{\cancel{y}}\,y\,y\,y\,y} = \dfrac{1}{y^4}$.

LAW 2

A. $\dfrac{x^m}{x^n} = x^{m-n}$ if m is bigger than n.

B. $\dfrac{x^m}{x^n} = \dfrac{1}{x^{n-m}}$ if n is bigger than m.

EXAMPLE 8—

$$\dfrac{30a^6b^7c^8}{2a^4b^2c} = 15a^2b^5c^7 \; (a^{6-4}b^{7-2}c^{8-1}) \text{ and } \dfrac{30}{2} = 15.$$

EXAMPLE 9—

$$\dfrac{a^3}{a^3} = \dfrac{8}{8} = \dfrac{3 \text{ pigs}}{3 \text{ pigs}} = 1.$$

EXAMPLE 10—

$$\dfrac{12a^9b^7c^5d^3e}{18a^4b^6c^5d^7e^5} = \dfrac{2a^5b}{3d^4e^4}; \; \dfrac{12}{18} = \dfrac{2}{3}, \; a^{9-4}, \; b^{7-6}, \; \dfrac{1}{d^{7-3}}, \; \dfrac{1}{e^{5-1}},$$

and $\dfrac{c^5}{c^5} = 1.$

Distributive Law

We end this chapter with perhaps the most favorite of the laws.

The distributive law: $a(b + c) = ab + ac$.

EXAMPLE 1—

$5(6x + 7y) = 30x + 35y.$

EXAMPLE 2—

$7(3b + 5c + f) = 21b + 35c + 7f.$

EXAMPLE 3

$4b^6(7b^5 + 3b^2 + 2b + 8) = 28b^{11} + 12b^8 + 8b^7 + 32b^6$.

EXAMPLE 4

Multiply and simplify:

$4(3x + 5) + 2(8x + 6) = 12x + 20 + 16x + 12 = 28x + 32$

EXAMPLE 5

Multiply and simplify:

$$5(2a + 5b) + 3(4a + 6b) = 10a + 25b + 12a + 18b$$
$$= 22a + 43b$$

EXAMPLE 6

Add and simplify:

$(5a + 7b) + (9a - 2b) = 5a + 7b + 9a - 2b = 14a + 5b$

Let us now learn about negative numbers!!

CHAPTER 2

INTEGERS PLUS

Later, you will probably look back at Chapter 1 as verrry easy. However it is new to many of you and may not seem easy at all. Relax. Most of Chapter 2 duplicates Chapter 1. The difference is that in Chapter 2 we will be dealing with integers.

The *integers* are the set . . . −3, −2, −1, 0, 1, 2, 3, . . .

Or written 0, ±1, ±2, ±3, . . .

The *positive integers* is another name for the natural numbers.

Nonnegative integers is another name for the whole numbers.

Even integers: 0, ±2, ±4, ±6, . . .

Odd integers: ±1, ±3, ±5, ±7, . . .

±8 means two numbers, +8 and −8.

x positive: $x > 0$, x is greater than zero.

x negative: $x < 0$, x is less than zero.

Don't know > or <? We will do later!

9 means +9.

EXAMPLE 1—

Graph the set −3, −2, 0, 1, 4.

$$-5\ -4\ -3\ -2\ -1\ \ 0\ \ 1\ \ 2\ \ 3\ \ 4\ \ 5$$

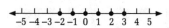

EXAMPLE 2—

Graph all the integers between −3 and 4.

EXAMPLE 3—

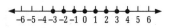

Graph all the integers between −3 and 4 inclusive.

Now that we know a little about integers, let's add, subtract, multiply, and divide them.

ADDITION

For addition, think about money. + means gain and − means loss.

EXAMPLE 4—

7 + 5.

Think (don't write!!) (+7) + (+5). Start at 7 and gain 5 more. Answer +12 or 12.

EXAMPLE 5—

−3 − 4.

Think (don't write) (−3) + (−4). Start at −3 and lose 4 more. Answer −7.

EXAMPLE 6—

−7 + 3.

Think (don't write) (−7) + (+3) Start at −7 and gain 3; lose 4. Answer −4.

EXAMPLE 7—

−6 + 9.

Think (don't write) (−6) + (+9).

Start at −6 and gain 9. We gained 3. Answer: +3 or 3.

NOTE

$-7 + 9$ is the same as $9 - 7 = (+9) + (-7) = 2$.

You should read these examples (and all the examples in the book) over until they make sense.

Here are the rules in words:

Addition 1: If two (or all) of the signs are the same, add the numbers without the sign, and put that sign.

Addition 2: If two signs are different, subtract the two numbers without the sign, and put the sign of the larger number without the sign.

EXAMPLE 8

$-7 -3 -2 -5 -1$.

All signs are negative; add them all, and put the sign. Answer: -18.

EXAMPLE 9

$-7 + 2$ or $2 - 7$.

Signs are different; subtract $7 - 2 = 5$. The larger number without the sign is 7. Its sign is $-$.

Answer: -5.

EXAMPLE 10

$7 - 3 - 9 + 2 - 8$. Add all the positives: $7 + 2 = 9$; add all the negatives: $-3 - 9 - 8 = -20$; then $9 - 20 = -11$.

EXAMPLE 11

Simplify $6a - 7b - 9a - 2b$.

Combine like terms: $6a - 9a = -3a$; $-7b - 2b = -9b$. Answer: $-3a - 9b$.

Just like the last chapter!!!!

SUBTRACTION

What we are doing is changing all subtraction problems to addition problems.

Definition: subtraction: $a - b = a + (-b)$.

Important: There are only two real subtraction problems.

$-7 - (-3) = (-7) + (+3) = -4$

or

$6 - (-2) = 6 + (+2) = 8$

A number followed by a minus sign followed by a number in parenthesis with a – sign in front of it.

$-2 - (+6) = -2 + (-6) = -8$

or

$7 - (+6) = 7 + (-6) = 1$

A number followed by a minus sign followed by a number in parenthesis with a + sign in front.

Only 1 sign between is always adding.

$-4 - (5)$ is the same as $-4 -5 = -9$.

$-3 + (-7)$ is also an adding problem; answer -10.

EXAMPLE 12—

Simplify $3x - (-7x) - (+4x) + (-9x) - 3x$.

$(3x) + (+7x) + (-4x) + (-9x) + (-3x) = 10x + (-16x) = -6x$

(At the beginning you might have to write out all the steps; your goal is to do as few steps as possible!)

MULTIPLICATION

Look at the pattern that justifies a negative number times a positive number gives a negative answer.

From 2 to 1 is down 1: $\begin{array}{l} (+2)(+3) = +6 \\ (+1)(+3) = +3 \end{array}$ Answer goes down 3.

From 1 to 0 is down 1: $\begin{array}{l}(+1)(+3) = +3\\(0)(+3) = 0\\(-1)(+3) = -3\end{array}$ Answer goes down 3.

From 0 to –1 is down 1: Answer goes down 3.

We have just shown a negative times a positive is a negative. (The same is true for division.) By the commutative law a positive times a negative is also a negative. Let's look at one more pattern.

From 2 to 1 is down 1: $\begin{array}{l}(+2)(-3) = -6\\(+1)(-3) = -3\\(0)(-3) = 0\\(-1)(-3) = +3\end{array}$ Answer goes up 3!!!!!

From 1 to 0 is down 1: From –3 to 0 is up 3.

From 0 to –1 is down 1: Answer is up 3.

What we just showed is a negative times a negative is a positive. (The same is true for division.)

More generally, we need to look at only negative signs in multiplication and division problems.

Odd number of negative signs, answer is **negative.**

Even number of negative signs, answer is **positive.**

EXAMPLE 13 (VERY IMPORTANT)—

A. -3^2; B. $(-3)^2$; C. $-(-3)^2$.

The answer in each case is 9. The only question is, "Is it +9 or –9?"

The answer is the number of minus signs.

A. $-3^2 = -(3 \times 3) = -9$ One minus sign

B. $(-3)^2 = (-3)(-3) = +9$ Two minus signs

C. $-(-3)^2 = -(-3)(-3) = -9$ Three minus signs

We need to explain a little.

A. The exponent is only attached to the number in front of it. -3^2 means negative 3^2.

B. If you want to raise a negative to a power, put a parenthesis around it: $(-3)^2$!

EXAMPLE 14—

$\dfrac{8(-1)(-4)(-20)}{(-16)(-2)(+2)}$. Five negative signs (odd number); answer is minus, -10.

EXAMPLE 15—

Remember: When you multiply, add the exponents if the base is the same!

$(-10a^3b^4c^5)(-2a^8b^9c^{100})(7a^2b^3c)$.

Determine the sign first: 2 − signs; answer is +. The rest of the numerical coefficient $(10)(2)(7) = 140$; $a^{3+8+2}b^{4+9+3}c^{5+100+1}$.

Answer: $+140a^{13}b^{16}c^{106}$. Notice, big numbers do NOT make hard problems.

EXAMPLE 16—

$\dfrac{(-3b^2)^4(-b^6)^3(4b)^3}{(2b^5)^5}$.

$\dfrac{(-3b^2)(-3b^2)(-3b^2)(-3b^2)(-b^6)(-b^6)(-b^6)(4b)(4b)(4b)}{(2b^5)(2b^5)(2b^5)(2b^5)(2b^5)}$

Seven − signs, an odd number. Answer is −.

Arithmetic Trick   Always divide (cancel) first. It makes the work shorter or much, much shorter and easier.

$$\dfrac{3 \times 3 \times 3 \times 3 \times \overset{1}{\cancel{4}} \times \overset{1}{\cancel{4}} \times \overset{2}{\cancel{4}}}{\cancel{2} \times 2 \times \cancel{2} \times \cancel{2} \times \cancel{2}} = 162$$

If you multiply first, on the top you would get $(3)(3)(3)(3)(4)(4)(4) = 5184$.

On the bottom you would get $(2)(2)(2)(2)(2) = 32$. Long dividing, you would get 162.

Canceling makes the problem much easier and much shorter.

Let's do the exponents:

Top: $(b^2)^4 = b^8$, $(b^6)^3 = b^{18}$, $b^3 = b^3$: $b^{8+18+3} = b^{29}$.

Bottom: $(b^5)^5 = b^{25}$.

$$\frac{b^{29}}{b^{25}} = b^4$$

Answer: $-162b^4$. (It takes longer to write out than to do.)

Remember: A power to a power, multiply the exponents.

Remember: If the base is the same, when you divide, you subtract exponents.

EXAMPLE 17—

$(-3a^2b^4)^3 = (-3)^3(a^2)^3(b^4)^3$ Answer: $-27a^6b^{12}$.

NOTE

Remember:
$(xy)^n = x^n y^n$.

EXAMPLE 18—

$\left(\dfrac{-2x^4}{3y^6}\right)^5$.

Top: $(-2)^5 = -32$; $(x^4)^5 = x^{20}$.

Bottom: $3^5 = 243$; $(y^6)^5 = y^{30}$.

Answer $\dfrac{-32x^{20}}{243y^{30}}$

Note: 5th and last law of exponents:
$$\left(\frac{a}{b}\right)^n = \frac{a^n}{b^n}$$

EXAMPLE 19—

$\left(\dfrac{12a^4b^5}{3a^9b^2}\right)^2$. Simplify inside ( ) first.

$= \left(\dfrac{12}{3}\ \dfrac{a^4}{a^9}\ \dfrac{b^5}{b^2}\right)^2$

$= \left(\dfrac{4b^3}{a^5}\right)^2 = \dfrac{16b^6}{a^{10}}$

EXAMPLE 20—

$\dfrac{(4a^3b^6)^2}{(2ab^3)^4}$. Outside exponents are different; they must be done first.

$= \dfrac{16a^6b^{12}}{16a^4b^{12}} = a^2$

RATIONAL NUMBERS

Informally, we call these numbers fractions, but this is not technically correct. There are two definitions of rational numbers:

Definition 1: Any number a/b where a and b are integers, b cannot be 0!

Definition 2: Any number that can be written as a repeating or terminating decimal.

$-\frac{5}{6} = -0.83333\ldots = -0.8\overline{3}$ is a rational number. So is $\frac{1}{4}$ $= .25\ (= .2500000\ldots = .250\overline{0})$. But $\pi/6$ is not a rational number (π is not rational) even though it is a "fraction."

What we call fractions are really rational numbers. For this book, even though technically incorrect, we may call them fractions.

EXAMPLE 21—

If $a = \frac{1}{2}$ and $b = -\frac{1}{4}$, evaluate $ab^2 - 6a - 2$.

$$ab^2 - 6a - 2 = \left(\frac{1}{2}\right)\left(\frac{-1}{4}\right)^2 - 6\left(\frac{1}{2}\right) - 2 = \left(\frac{1}{2}\right)\left(\frac{1}{16}\right)$$

$$- 6\left(\frac{1}{2}\right) - 2 = \frac{1}{32} - 3 - 2 = -5 + \frac{1}{32} = \frac{-159}{32} \text{ or } -4\frac{31}{32}$$

EXAMPLE 22—

Simplify $\dfrac{3}{8}a - \dfrac{5}{6}b - \dfrac{1}{16}a + a - \dfrac{1}{3}b$.

Answer: $\dfrac{3}{8}a - \dfrac{1}{16}a + a - \dfrac{5}{6}b - \dfrac{1}{3}b = \dfrac{6}{16}a + \dfrac{16}{16}a - \dfrac{1}{16}a$

$$- \frac{5}{6}b - \frac{2}{6}b = \frac{21}{16}a - \frac{7}{6}b.$$

If you are pretty much OK until the last two examples, then your problem is fractions. Please read the next chapter carefully and then practice your fractions!!!!!

DIVISION

We have already had the facts that $\dfrac{x^{11}}{x^2} = x^9$ and

$\dfrac{y^3}{y^{99}} = \dfrac{1}{y^{96}}$.

To summarize, let's do a few more examples.

A. $\dfrac{18a^5 b^6 c^7 d^8 e^9}{9a^8 b^3 c^7 d^9 e} = \dfrac{2b^3 e^8}{a^3 d}$.

$\dfrac{18}{9} = 2 \qquad \dfrac{a^5}{a^8} = \dfrac{1}{a^3} \qquad \dfrac{b^6}{b^3} = b^3 \qquad \dfrac{d^8}{d^9} = \dfrac{1}{d} \qquad \dfrac{e^9}{e} = e^8$

and $\dfrac{c^7}{c^7} = 1$

Notice, long is not hard!!

B. $\dfrac{4a^8b^5}{10a^2b^9} = \dfrac{2a^6}{5b^4}$. 4/10 is reduced to 2/5. Remember reducing is division.

C. $\dfrac{2a^2}{6a^6} = \dfrac{1}{3a^4}$.

D. $\dfrac{11^5}{11^{20}} = \dfrac{1}{11^{15}}$. Exponents are subtracted; base stays the same.

Short Division

This is the opposite of adding fractions, which says

$$\frac{a}{c} + \frac{b}{c} = \frac{a+b}{c} \quad \text{orrr} \quad \frac{2}{7} + \frac{3}{7} = \frac{5}{7}.$$

The opposite is $\dfrac{a+b}{c} = \dfrac{a}{c} + \dfrac{b}{c}$.

EXAMPLE 23—

$$\frac{55a^9 + 35a^7}{5a^5} = \frac{55a^9}{5a^5} + \frac{35a^7}{5a^5} = 11a^4 + 7a^2.$$

EXAMPLE 24—

$$\frac{16a^8 - 18a^7 + 4a^3 + 2a}{4a^3}.$$

$$\frac{16a^8}{4a^3} - \frac{18a^7}{4a^3} + \frac{4a^3}{4a^3} + \frac{2a}{4a^3} = 4a^5 - \frac{9a^4}{2} + 1 + \frac{1}{2a^2}$$

1. The "1" must be the third term because all the terms are added (anything except 0 over itself is 1).

2. If there are four unlike terms in the top of the fraction at the start, there must be four terms in the answer.

3. There is another way to do this problem, which we will do soon.

4. Most students like this kind of problem, but **many** forget how to do it. Please, don't forget.

DISTRIBUTIVE LAW REVISITED

Let's do the distributive law, with minus signs now.

$$a(b + c) = ab + ac \qquad \text{and} \qquad a(b - c) = ab - ac$$

EXAMPLE 25—

Question	Answer
A. $7(3a - 4b + 5c)$	$21a - 28b + 35c$
B. $5a(3a - 4b + 9c)$	$15a^2 - 20ab + 45ac$ (Letters written alphabetically look prettier.)
C. $6m^2(9m^3 - m + 3)$	$54m^5 - 6m^3 + 18m^2$
D. $-4x^3y^4(-2x^5y^2 + 3xy^5)$	$+8x^8y^6 - 12x^4y^9$

EXAMPLE 26—

Multiply and simplify.

Question	Answer
A. $5(2a - 3b) + 7(4a + 6b)$	$10a - 15b + 28a + 42b = 38a + 27b$
B. $2(3x - 4) - 5(6x - 7)$	$6x - 8 - 30x + 35 = -24x + 27; (-5)(-7)$ $= +35$, careful!!!
C. $2c(4c - 7) + 5c(2c - 1)$	$8c^2 - 14c + 10c^2 - 5c = 18c^2 - 19c$
D. $(2x + y - z) + (5z - y) - (4y - 2x)$	$1(2x + y - z) + 1(5z - y) - 1(4y - 2x)$ $= 2x + y - z + 5z - y - 4y + 2x = 4x - 4y$ $+ 4z$

FACTORING

We would like to do the reverse of the distributive law, taking out the greatest common factor. It is the start of a wonderful game I love. It is a very, very important game. The rest will be found in *Algebra for the Clueless*. But first . . .

We are looking for the GREATEST COMMON FACTOR.

There are three words: factor, common, greatest.

EXAMPLE 27—

Find the greatest common factor of 24 and 36.

Factors of 24: 1, 2, 3, 4, 6, 8, 12, 24. Factors of 36: 1, 2, 3, 4, 6, 12, 18, 36.

Common factors of 24 and 36: 1, 2, 3, 4, 6, 12.

The GREATEST common factor is 12.

EXAMPLE 28—

Find the greatest common factor:

Question	Answer
A. $6x$, $9y$	3
B. $8abc$, $12afg$	$4a$
C. $10a^4$, $15a^7$	$5a^4$ (the lowest power of an exponent, if the letter occurs in all terms)

Taking Out the Greatest Common Factor

The distributive law says $3(4x + 5) = 12x + 15$.

Given $12x + 15$. Factoring says 3 is the greatest common factor. 3 times what is $12x$ (or you could say $12x/3$). Answer is $4x$. 3 times what is 15. Answer is 5. $12x + 15 = 3(4x + 5)$.

EXAMPLE 29—

Factor completely:

Question	Answer
A. $8x - 12y$	$4(2x - 3y)$
B. $10ab - 15b^2 - 25bc$	$5b(2a - 3b - 5c)$
C. $27a^6b^7c^8 + 36a^{80}b^2c^{11}d^5 + 18a^{1000}b^5cd^4$	$9a^6b^2c(3b^5c^7 + 4a^{74}c^{10}d^5 + 2a^{994}b^3d^4)$

NOTES FOR C

1. When the only factor of the numbers in ( ) is 1, you've taken out the largest numerical common factor.

2. a^6 and b^2 are the lowest exponent in common. c is to the first power in the third term. All terms do NOT have a "d." So no d's can be factored out.

3. Check? Multiply out using the distributive law to see if you get the beginning.

4. Large numbers do NOT make a hard problem.

5. No matter how many kinds of factoring you will learn in the future, this is always the first you do.

6. You must practice this (and future factoring skills). Not knowing factoring is a major reason why some students can't go on. Let us do two more.

Question	Answer
D. $ab^3 + a^2b + ab$	$ab(b^2 + a + 1)$ (Three terms in the original problem, three terms inside since $1(ab) = ab$.)
E. $6ab + 8ac + 9bc$	2 is a factor of 6 and 8 but not 9; 3 is a factor of 6 and 9 but not 8; a, b, and c are in two of the terms but not all. This is a PRIME! It can't be factored.

SHORT DIVISION REVISITED

EXAMPLE 30—

$\dfrac{8x^6 + 12x^5}{2x^3}$. Factor the top: $\dfrac{4x^5(2x + 3)}{2x^3} = 2x^2(2x + 3)$.

$\left(\dfrac{4x^5}{2x^3} = 2x^{5-3} = 2x^2\right)$.

Let's do a thorough study of fractions. If you absolutely know it, you can skip it.

CHAPTER 3

FRACTIONS, WITH A TASTE OF DECIMALS

This is a vital chapter. It probably should be Chapter 1. But if I made this chapter the first chapter, you probably wouldn't have bought the book.

The first key of math is to know your multiplication tables absolutely perfectly and instantly. You must. Next you should be able to add, subtract, multiply, and short divide whole numbers well, without a calculator. Most people who do not know how to do this tend to lose their confidence in doing higher-level math.

ENOUGH TALKING.
I TALK *TOO* MUCH!
LET'S DO FRACTIONS.

Fractions (the Positive Kind)

One of the big problems about fractions is that many people do not understand what a fraction is.

Suppose I am 6 years old. I ask you, "What is ⅔?" Oh, I am a smart 6-year-old. I'll give you a start.

Suppose I have a pizza pie. . . . We divide it into seven equal parts, and we have three of them. That is ³⁄₇. Now which is bigger, ²⁄₇ or ³⁄₇? ²⁄₇ means you have 2 pieces out of 7, and ³⁄₇ means that you get 3 pieces out of 7. Since 3 pieces is more than 2, ³⁄₇ is bigger than ²⁄₇.

Which is bigger, ³⁄₇ or ³⁄₈? We have the same number of pieces, 3. If we divide a pie into 7 pieces, the pieces are larger than if we divide the pie into 8 parts. So ³⁄₇ is bigger than ³⁄₈.

We get the following rules: If the denominators (bottoms) are the same, the bigger the numerator (top), the bigger the fraction. If the numerators are the same, the bigger the denominators, the smaller the fraction.

NOTE

Just for completeness, negatives are the opposite. If 2 is less than 3, but −2 is greater than −3 (if you are losing 2 dollars, it is not as bad as if you losing 3 dollars). Similarly, ²⁄₇ is less than ³⁄₇, but −²⁄₇ is greater than −³⁄₇.

When is a fraction equal to 1? When the top equals the bottom. See the pictures ⁵⁄₅ = ⁷⁄₇ = . . . = 1 whole.

When is a fraction more than 1? When the top is bigger than the bottom. See the picture that shows ³⁄₂ and ⁷⁄₄ among many others are bigger than 1.

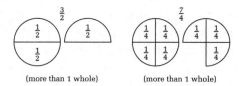

(more than 1 whole)   (more than 1 whole)

You should learn dividing and multiplying fractions before you add and subtract them. But first . . .

REDUCING

METHOD 1 Suppose we want to reduce a fraction. We find the largest number to divide into both the top and bottom, the greatest common factor.

METHOD 2 We write numerator and denominator as the product of primes and cancel those in common.
  Reduce ⁶⁄₉.

METHOD 1 The largest common factor of 6 and 9 is 3. 3 into 6 is 2. 3 into 9 is 3. ⁶⁄₉ = ²⁄₃.

Reason Take a look at the pictures. ⁶⁄₉ of the pie is the same as ²⁄₃ of the pie. The only difference is in ⁶⁄₉ the pieces are divided into parts.

METHOD 2 $6 = 3 \times 2$ and 9 is 3×3. $\dfrac{6}{9} = \dfrac{2 \times \cancel{3}^{1}}{3 \times \cancel{3}_{1}} = \dfrac{2}{3}$.

Why can we do *this?* We can because we can always cancel factors.
  When you divide (or multiply) by 1, everything stays the same.

²⁄₃   ⁶⁄₉

EXAMPLE 2—

Reduce $^{48}\!/_{60}$.

METHOD 1 It is not necessary to know the largest common factor at once although it would save time. You might see 4 is a common factor. $^{48}\!/_{60} = {}^{12}\!/_{15}$. Now you see 3 is a common factor. $^{12}\!/_{15} = {}^{4}\!/_{5}$, the answer.

METHOD 2 Write 48 and 60 as the product of primes.

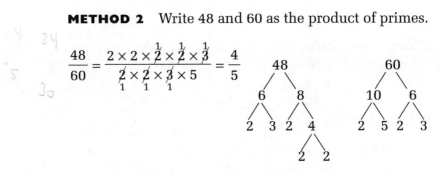

$$\frac{48}{60} = \frac{2 \times 2 \times \overset{1}{\cancel{2}} \times \overset{1}{\cancel{2}} \times \overset{1}{\cancel{3}}}{\underset{1}{\cancel{2}} \times \underset{1}{\cancel{2}} \times \underset{1}{\cancel{3}} \times 5} = \frac{4}{5}$$

Method 2 is not the shortest method, but is necessary for two reasons. First, when numbers are large, it is the most efficient way. Much more important, the second is the way to determine common denominators when we add fractions. We might do a little in this book, but we do a lot in *Algebra for the Clueless*.

MULTIPLICATION AND DIVISION

RULE $\dfrac{a}{b} \times \dfrac{c}{d} = \dfrac{ac}{bd}$ In multiplication, we multiply the tops, and multiply the bottoms.

EXAMPLE 1—

$\dfrac{2}{3} \times \dfrac{5}{7} = \dfrac{10}{21}$. Why? You must always understand why.

Let us draw a picture. $^{5}\!/_{7}$ looks like this . . .

⅓ of ⁵⁄₇ looks like this . . .

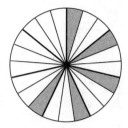

⅔ of ⁵⁄₇, twice ⅓ of ⁵⁄₇ looks like this. . . . Count!

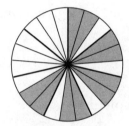

The pie is cut into 21 parts. You have 10 of them, ¹⁰⁄₂₁.

EXAMPLE 2—

Multiply $\dfrac{4}{15} \times \dfrac{25}{6}$.

$$\frac{4}{15} \times \frac{25}{6} = \frac{\overset{2}{\cancel{4}}}{15} \times \frac{25}{\underset{3}{\cancel{6}}} = \frac{2}{15} \times \frac{\overset{5}{\cancel{25}}}{3} = \frac{2}{3} \times \frac{5}{3} = \frac{10}{9}$$

or $\dfrac{4}{15} \times \dfrac{25}{6} = \dfrac{\overset{2}{\cancel{4}}}{\underset{3}{\cancel{15}}} \times \dfrac{\overset{5}{\cancel{25}}}{\underset{3}{\cancel{6}}} = \dfrac{2}{3} \times \dfrac{5}{3} = \dfrac{10}{9}$

NOTES

1. Reduce (cancel) before you multiply.
2. You can't cancel two tops or two bottoms.
3. You should do all canceling in one step.

EXAMPLE 3—

Multiply $\dfrac{8}{9} \times \dfrac{15}{100}$.

$$\frac{8}{9} \times \frac{15}{100} = \frac{\overset{2}{\cancel{8}}}{\underset{3}{\cancel{9}}} \times \frac{\overset{5}{\cancel{15}}}{\underset{25}{\cancel{100}}} = \frac{2}{3} \times \frac{\overset{1}{\cancel{5}}}{\underset{5}{\cancel{25}}} = \frac{2}{3} \times \frac{1}{5} = \frac{2}{15}$$

4. You may cancel the top and the bottom of the same fraction.

EXAMPLE 4

Multiply $4 \times \dfrac{3}{125}$. $4 \times \dfrac{3}{125} = \dfrac{4}{1} \times \dfrac{3}{125} = \dfrac{12}{125}$. Now let's divide.

DIVISION RULE $\dfrac{a}{b} \div \dfrac{c}{d} = \dfrac{a}{b} \times \dfrac{d}{c} = \dfrac{a \times d}{b \times c} = \dfrac{ad}{bc}$.

Invert (flip upside down) the second fraction and multiply, canceling if necessary.

Okay why is this true? Let's give a simple example to show it.

EXAMPLE 5

A. $7 \div 2$. This is pictured below.

B. $7 \div 2 = \dfrac{7}{1} \div \dfrac{2}{1} = \dfrac{7}{1} \times \dfrac{1}{2} = \dfrac{7}{2}$.

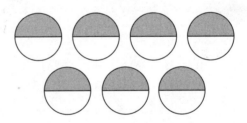

In words, 7 divided by 2 is half of 7 or 7 times one-half.

We can see that it is the same as (A) which is $3\dfrac{1}{2}$.

EXAMPLE 6

$\dfrac{4}{11} \div \dfrac{10}{7} = \dfrac{4}{11} \times \dfrac{7}{\overset{}{\underset{5}{10}}} = \dfrac{2}{11} \times \dfrac{7}{5} = \dfrac{14}{55}$.

ADDING AND SUBTRACTING

Adding and subtracting are done in exactly the same way. So we'll only do addition.

RULE If the denominators are the same, add or subtract the tops. In symbols, $\dfrac{a}{c} \pm \dfrac{b}{c} = \dfrac{a \pm b}{c}$.

EXAMPLE I—

$$\frac{3}{11} + \frac{9}{11} = \frac{12}{11}; \frac{7}{99} - \frac{5}{99} = \frac{2}{99}.$$

$\pm$ **means + or −.**

The problem occurs when the bottoms are *not* the same. We must find the *least common denominator,* which is really the least *common multiple,* LCM.

For example, find the least common multiple of 3 and 4.

The LCM consists of three words: least, common, multiple.

Multiples of 3 $(3 \times 1, 3 \times 2, 3 \times 3, \ldots)$; 3, 6, 9, 12, 15, 18, 21, 24, 27, 30, 33, 36, 39, 42, 45, 48, 51, . . .

Multiples of 4 4, 8, 12, 16, 20, 24, 28, 32, 36, 40, 44, 48, 52, . . .

Common **Multiples** 12, 24, 36, 48, . . .

Least **Common Multiple** 12.

There are three techniques for adding (subtracting) fractions that have unlike denominators: one for small denominators, one for medium denominators, one for large denominators.

EXAMPLE 2—

LCD is 12, which you
should see.*

$$\frac{1}{4} + \frac{5}{6} = \frac{3}{12} + \frac{10}{12} = \frac{13}{12}.$$

4 into 12 is 3. 1 × 3 = 3. 6
in 12 is 2. 2 × 5 = 10.

Another way to look at this problem is the following:

* If you cannot see it,
look at the next problem.

$$\frac{1}{4} + \frac{5}{6} = ?????$$

$$\frac{1}{4} = \frac{1 \times 3}{4 \times 3} = \frac{3}{12}$$

$$\frac{5}{6} = \frac{5 \times 2}{6 \times 2} = \frac{10}{12}$$

and $\dfrac{3}{12} + \dfrac{10}{12} = \dfrac{13}{12}$

We are allowed to do this because multiplying by ¾ or ½ is multiplying by 1. When you multiply by 1, the expression has the same value.

This technique is very important, first, for fractions with BIG denominators. More important, it is the way you add algebraic fractions found in *Algebra for the Clueless.*

ADDING MEDIUM DENOMINATORS

EXAMPLE—

Add: ¾ + ⅚ + ⅛ + ⁴⁄₉. What is the LCM? It is hard to tell, although some of you may be able.

The technique is to find multiples of the largest denominator.

9, 18, 27, 36, 45, 54, 63, 72, 81, 90, 99, . . . We see that 72 is the first number that 4, 6, and 8 are also factors.

$$\frac{3(18)}{4(18)} = \frac{54}{72}$$

$$\frac{5(12)}{6(12)} = \frac{60}{72}$$

$$\frac{1(9)}{8(9)} = \frac{9}{72}$$

$$\frac{4(8)}{9(8)} = \frac{32}{72}$$

Total is $\dfrac{155}{72}$ or $2\,\dfrac{11}{72}$. Not bad so far.

ADDING LARGE DENOMINATORS

EXAMPLE—

Add: $\dfrac{9}{100} + \dfrac{25}{48} + \dfrac{5}{108} + \dfrac{1}{30}$.

What is the LCM (LCD) of 100, 48, 108, and 30? You should know. . . . I'm just fooling. Almost no one, including me, knows the answer. This is how we do it.

STEP 1 Factor all denominators into primes:

$100 = 2 \times 2 \times (5 \times 5)$

$48 = (2 \times 2 \times 2 \times 2) \times 3$

$108 = 2 \times 2 \times (3 \times 3 \times 3)$

$30 = 2 \times 3 \times 5$

STEP 2 The *magical* phrase is that the LCD (LCM) is the product of the most number of times a prime appears in any ONE denominator.

2 appears twice in 100, 4 times in 48, twice in 108, and once in 30. Four 2s are necessary.

3 appears not at all in 100, once in 48, 3 times in 108, and once in 30. Three 3s are needed.

We need two 5s, since it appears the most number of times, twice, in 100. The LCD (LCM) = $2 \times 2 \times 2 \times 2 \times 3 \times 3 \times 3 \times 5 \times 5$ Whew!!!

STEP 3   Multiply top and bottom by *what's missing* (missing factors in the LCD). (Multiplying by 1 keeps the value the same.)

Let's take a look at one of them, $\frac{5}{108}$.

108 has two 2s; LCD has four 2s; 2×2 is missing. 108 has three 3s; LCD has three 3s. No 3s are missing.

108 has zero 5s; LCD has two 5s; 5×5 is missing. Multiply top and bottom by what's missing!

$$\frac{5}{108} = \frac{5}{2 \times 2 \times 3 \times 3 \times 3}$$

$$= \frac{5 \times (2 \times 2 \times 5 \times 5)}{2 \times 2 \times 3 \times 3 \times 3 \times (2 \times 2 \times 5 \times 5)}$$

$$= \frac{5 \times 2 \times 2 \times 5 \times 5}{2 \times 2 \times 2 \times 2 \times 3 \times 3 \times 3 \times 5 \times 5}$$

Do the same for each fraction.

Unfortunately the next step with numbers is much messier than with letters. You must multiply out all tops and bottoms.

$$\frac{9}{100} = \frac{9}{2 \times 2 \times 5 \times 5}$$

$$= \frac{9 \times (2 \times 2 \times 3 \times 3 \times 3)}{2 \times 2 \times 5 \times 5 \times (2 \times 2 \times 3 \times 3 \times 3)} = \frac{972}{10,800} \quad 3 \times$$

$$\frac{25}{48} = \frac{25}{2 \times 2 \times 2 \times 2 \times 3}$$

$$= \frac{25 \times (3 \times 3 \times 5 \times 5)}{2 \times 2 \times 2 \times 3 \times (3 \times 3 \times 5 \times 5)} = \frac{5625}{10,800}$$

$$\frac{5}{108} = \frac{5}{2 \times 2 \times 3 \times 3 \times 3}$$

$$= \frac{5 \times (2 \times 2 \times 5 \times 5)}{2 \times 2 \times 3 \times 3 \times 3 \times (2 \times 2 \times 5 \times 5)} = \frac{500}{10,800}$$

$$\frac{1}{30} = \frac{1}{2 \times 3 \times 5}$$

$$= \frac{1 \times (2 \times 2 \times 2 \times 3 \times 3 \times 5)}{2 \times 3 \times 5 \times (2 \times 2 \times 2 \times 3 \times 3 \times 5)} = \frac{360}{10,800}$$

LAST STEPS   Add the tops and reduce.
If we add, we get 7457 / 10,800.

Now we have to reduce it. No, no no, I'm not kidding. It is not nearly as bad as it looks. 10,800 has only three prime factors: 2, 3, and 5. 7457 is not divisible by 2 since the last number is odd. 7457 is not divisible by 5 since it doesn't end in a 0 or a 5.

There is a trick for 3 (same for 9): 7 + 4 + 5 + 7 = 23 is not divisible by 3, so neither is 7457.

$\dfrac{7,457}{10,800}$ is the final answer.

DECIMALS

Of percentages, fractions, and decimals, probably the easiest is decimals, except possibly reading them. The secret is to read the whole part, say "and" for the decimal point, read the number as if it were a whole number and say the last decimal place.

EXAMPLE 1—

.47   47 hundredths

EXAMPLE 2—

.002   2 thousandths

EXAMPLE 3—

.1234   1234 ten thousandths

EXAMPLE 4—

23.06   23 and 6 hundredths

Let's put down one big number.

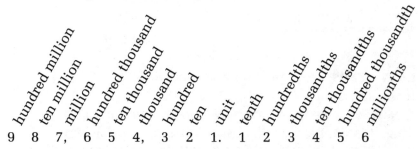

9 8 7, 6 5 4, 3 2 1. 1 2 3 4 5 6

Nine hundred eighty-seven million, six hundred fifty-four thousands, three hundred twenty-one AND one hundred twenty-three thousand, four hundred fifty-six millionths.

NOTE

You only say the word *and* when you read the decimal point. Listen to most TV people (except on *Jeopardy,* where people have written in). When people read numbers like 4,000,001, they will say 4 million and 1 when it should be 4 million, 1. How about 2002? Two thousand, two!!!

Decimals $(+, -, \times, \div)$

The rule for adding and subtracting is that you must line up the decimal point.

EXAMPLE 1—

$72.3 + 41 + 3.456$. 41 is a whole number. $41 = 41$.

$$
\begin{array}{r}
72.3 \\
41. \\
3.456 \\
\hline
116.756
\end{array}
$$

EXAMPLE 2—

$24.6 - .78$.

$$
\begin{array}{r}
24.6 \\
-.78
\end{array}
\quad = \quad
\begin{array}{r}
24.60 \\
-0.78 \\
\hline
23.82
\end{array}
$$

EXAMPLE 3—

$67.89 \times .876$.

Multiply and get 5947164. 67.89 has 2 numbers to the right of the decimal point; .876 has 3 numbers to the right of the decimal point; the answer must have five numbers to the right of the decimal point. Final answer is 59.47164.

Did you ever want to know why you add the number of places? Here's the reason.

If the first number is to the hundredths place and the second number is to the thousandths place, and you multiply $\frac{1}{100} \times \frac{1}{1,000}$, you get $\frac{1}{100,000}$, five places. That's all!!!!

EXAMPLE 4—

Divide 123.42 by .004.

$$.004\overline{)123.42} = \mathbf{.004.}\overline{)123.\mathbf{420.}}$$

Answer is 30,855.

Did you ever want to know why we want to divide by a whole number? It's easier!!!!

$$\frac{123.42}{.004} \times \frac{1,000}{1,000} = \frac{123,420}{4} = 30,855$$

Multiply the bottom by 1000—three places to the right; you must multiply the top by 1000 since 1000/1000 = 1 keeps the fraction the same value. That's all.

Decimals to Fractions and Fractions to Decimals

Let's finish up . . .

Decimals to fractions: "Read it and write it."

123.47 One hundred twenty-three and forty-seven hundredths: $123 \frac{47}{100}$. That's it.

Fractions to decimals. Divide the bottom into the top, decimal point 4 zeros.

$$\frac{3}{8} = 8\overline{)3.0000} = .375 \qquad \frac{1}{6} = 6\overline{)1.0000} = .16666\ldots = .1\overline{6}$$

We'll do percentages later. Now for an algebraic chapter that most students really like.

CHAPTER 4

FIRST-DEGREE EQUATIONS

This chapter is one of the favorites of most students. Most students have had equations as early as the first grade. We wrote $\square + 3 = 5$. This was an equation. We will be a little more formal. Imagine a balance scale. The equal sign is in the middle.

$$3xy - 4z^3 = 12x - 7y - x^{20} + 7$$
Left side =   Right side

We, of course, are going to work with much simpler equations!!!!

You should imagine a balance scale. You can (mathematically) rearrange either side (use the distributive law, combine like terms, . . .), and everything is OK. But if you add something to the left, you must add the same thing to the right in order to keep the scale balanced. The same goes for subtract, multiply, and divide.

A *solution* (root) of an equation is the number or expression that balances the sides.

EXAMPLE I—

Show 7 is not a root but 5 is a root of the equation $8x - 6 = 6x + 4$.

$$8(7) - 6 = 6(7) + 4 \qquad\qquad 8(5) - 6 = 6(5) + 4$$

$$50 \neq 46 \qquad\qquad\qquad 34 = 34$$

No, 7 is NOT a root.   Yes, 5 is a root to this equation.

EXAMPLE 2—

Show $7x - 2x - 3a = 2x + 24a$ has a root $x = 9a$

$$7(9a) - 2(9a) - 3a = 2(9a) + 24a$$

$$42a = 42a$$

Yes, $9a$ is a root.

There are three kinds of basic equations.

1. CONTRADICTION: an equation that has no solution or is false.

EXAMPLE 3—

A. $9 + 7 = 4$.

B. $x = x + 1$. (There is no number that when you add 1 to it gives you the same number.)

2. IDENTITY: an equation that is always true as long as it's defined.

EXAMPLE 4—

A. $5x + 9x = 14x$.

No matter what value x is, five times the number plus nine times the number is fourteen times the number.

B. $\dfrac{3}{x} + \dfrac{2}{x} = \dfrac{5}{x}$. It is not defined if $x = 0$, because

division by zero is undefined!!! Otherwise it is always true.

You will have to wait for trig and *Precalc with Trig for the Clueless* to do these.

NOTE

If there are two letters in a problem, you must be told what letter to solve for.

3. CONDITIONAL EQUATION: an equation that is true for some values but not all values.

Examples 1 and 2 are conditional equations. These linear or first-degree equations have one answer.

A QUADRATIC EQUATION, such as $x^2 - 2x = 8$, is also a conditional equation. It has two roots, $x = 4$ and $x = -2$. $(4)^2 - 2(4) = 8$ and $(-2)^2 - 2(-2) = 8$. You can see them in *Algebra for the Clueless.* Let us solve some linear equations.

SOLVING LINEAR EQUATIONS

There are four one-step equations.

EXAMPLE 5

Solve for x:

$x + 9 = 11$

$x + 9 = 11$

$\underline{-9 = -9}$

$x + 0 = 2$

$x = 2$

Check: $2 + 9 = 11$.

We always solve for one of the unknowns.

To get rid of an addition you subtract or add a negative; the same to each side!! We will not put in the 0 in the future, Try to keep the equal signs lined up.

EXAMPLE 6

$x - 9 = -20$

$\underline{+9 = +9}$

$x = -11$

Check: $-11 - 9 = -20$.

To get rid of a subtraction (or adding a negative) you add (add a positive).

EXAMPLE 7—

To get rid of a multiplica-tion, you divide.

$$4x = 20$$

$$\frac{4x}{4} = \frac{20}{4}$$

$$x = 5$$

Check: $4(5) = 20$.

EXAMPLE 8—

$$\frac{x}{3} = 7$$

To get rid of a division, you multiply.

$$\frac{3}{1} \cdot \frac{x}{3} = 3(7)$$

$$x = 21$$

Solving Linear Equations (in General)

It is quite hot outside, a good day to write a section most of you will like.

EXAMPLE 9—

Opposites (additive inverse):
$a + (-a) = (-a) + a = 0.$

Solve for x: $5x - 11 = 15x + 49$.

The opposite of 3 is –3 since $3 + (-3) = (-3) + 3 = 0.$

We would like to list the steps that will allow us to solve equations easily.

The opposite of –6a is 6a since $(-6a) + 6a = 6a + (-6a) = 0.$

The opposite of $0 = 0$.

1. Multiply each term by the LCD.

    1. **Fractions cause the most problems; get rid of them as soon as possible.**

2. If the x appears only on the right side, switch the sides.

3. Multiply out all parentheses, brackets, and braces.

4. On each side, combine like terms.

5. Add the opposite of the x terms on the right to each side.

6. Add the opposite of the non-x terms on the left to each side.

    2. **The purpose is, at the start, to make all equations look the same (so you can get very good very fast). The reason you are allowed to reverse sides is because of the SYMMETRIC LAW which says if $a = b$, then $b = a$.**

7. Factor out the x.

    7. **This step occurs only if there are two or more letters.**

8. Divide each side by the whole coefficient of x, including the sign.

Let's rewrite the equation and solve it.

1. **No fractions.**

2. **x terms are on both sides.**

Example 9 revisited.

3. **No parenthesis.**

4. **No like terms on each side: left side $5x - 11$, no like terms; $15x + 49$, no like terms.**

$$5x - 11 = 15x + 49$$

$$\underline{-15x} \quad = \underline{-15x} \qquad \text{(step 5)}$$

5. **Add the opposite of the x term on the right to each side; the opposite of $15x = -15x$.**

6. Add the opposite of the non-x term on the left to each side; the opposite of -11 is 11.

$$-10x - 11 = 49 \qquad \text{(step 6)}$$

$$\underline{+11 = 11}$$

$$-10x = 60$$

7. Only one letter.

8. Divide each side by the whole coefficient of x, which is -10.

$$\frac{-10x}{-10} = \frac{60}{-10} \qquad \text{(step 8)}$$

$$x = -6$$

Check: $5(-6) - 11 = 15(-6) + 49$. $-41 = -41$. Yes, -6 is correct!!!!!!! Always check in the original equation!!

If you follow these problems, you will not have to memorize the steps. You will find that you will learn them. Try these problems before you try others in a book with problems. Make sure you understand each step.

No step 1 or step 2 or step 3

EXAMPLE 10—

Solve for y $4y + 7y + 3 = 2y + 5 + 25$.

$$4y + 7y + 3 = 2y + 5 + 25$$

4. Combine like terms on each side: left side: $4y + 7y = 11y$; right side: $5 + 25 = 30$.

$$11y + 3 = 2y + 30$$

5. Add $-2y$ to each side.

$$\underline{-2y \qquad = -2y}$$

$$9y + 3 = 30$$

6. Add -3 to each side.

$$-3 = -3$$

7. Divide each side by 9.

$$9y = 27$$

$$y = 3$$

Check: $4(3) + 7(3) + 3 = 2(3) + 5 + 25$. $36 = 36$, and the answer checks!!

EXAMPLE 11—

Solve for x.

$$2(3x - 4) - 4(2x - 5) = 6(2 - x)$$

No step 1 or step 2.

3. Multiply out (). Be careful with the second parenthesis: $-4(2x - 5) = -8x + 20$.

$$6x - 8 - 8x + 20 = 12 - 6x$$

$$-2x + 12 = -6x + 12$$

4. Combine like terms on the left.

$$\underline{+6x \qquad = +6x}$$

5. Add $6x$ to both sides.

$$4x + 12 = 12$$

$$\underline{-12 = -12}$$

6. Add -12 to each side.
7. No step 7.

$$4x = 0$$

$$x = 0$$

8. Divide each side by 4; $4x/4 = x$; $0/4 = 0$.

Check: $2(3(0) - 4) - 4(2(0) - 5) = 6(2 - 0)$; $-8 + 20 = 12$.

EXAMPLE 12—

Solve for x.

$$\frac{5}{6} = \frac{x}{3} + \frac{x + 1}{4}$$

$$\frac{12}{1} \cdot \frac{5}{6} = \frac{12}{1} \cdot \frac{x}{3} + \frac{12}{1} \cdot \frac{(x + 1)}{4}$$

$$10 = 4x + 3(x + 1)$$

1. Multiply each term by LCD: $12 = 12/1$. If you have a fraction with more than one term on top, such as $x + 1$, for safety, put a parenthesis around it.

Look how much nicer this is without fractions.

2. x terms are only on the right; switch sides.

$$4x + 3(x + 1) = 10$$

3. Multiply out parenthesis on the left.

4. Combine like terms on the left.

$$4x + 3x + 3 = 10$$

5. No x terms on the right.

$$7x + 3 = 10$$

6. Add -3 to each side. No step 7.

$$-3 = -3$$

8. Divide each side by 7.

$$7x = 7$$

$$x = 1$$

Check: $\frac{5}{6} = \frac{1}{3} + \frac{1}{2}$, since $\frac{2}{6} + \frac{3}{6} = \frac{5}{6}$.

EXAMPLE 13—

1. Write y as $y/1$. With two fractions, you usually cross-multiply $a/b = c/d$ means $ad = bc$.

Solve for x.

$$y = \frac{3x + 5}{2x - 7}$$

$3/4 = 6/8$ since $3(8) = 4(6)$. No step 2 (x is on both sides).

$$\frac{y}{1} = \frac{(3x + 5)}{(2x - 7)}$$

3. Multiply out parentheses.

$$y(2x - 7) = 1(3x + 5)$$

$$2xy - 7y = 3x + 5$$

No step 4. (No like terms on either side.)

5. Add $-3x$ to each side. Notice $2xy$ and $-3x$ are NOT, NOT, NOT like terms.

$$-3x = -3x$$

$$2xy - 3x - 7y = 5$$

6. Add the opposite of the non-x term(s) on left to each side. Add $+7y$ to each side.

$$+7y = +7y$$

$$2xy - 3x = 7y + 5$$

$$x(2y - 3) = 7y + 5$$

7. Factor out the x.

$$\frac{x(2y-3)}{(2y-3)} = \frac{7y+5}{2y-3}$$

$$x = \frac{7y+5}{2y-3}$$

8. **Divide each side by the whole coefficient of x; this is $2y - 3$.**

Notice   Although two letters seems messier, there is little or no arithmetic in the problems.

Let us try a few problems with words.

PROBLEMS WITH WORDS

We will do a few problems at this point and add some more as we learn more.

EXAMPLE 1—

Four more than five times a number is the same as ten less than seven times the number.

Find the Number   Let $x =$ the number. Four more than five times the number is $4 + 5x$ or $5x + 4$. Ten less than seven times the number is $7x - 10$. "Is" (was, will be) is usually the equal sign. It is here!

$$5x + 4 = 7x - 10$$

$$\underline{-7x \qquad = -7x}$$

$$-2x + 4 = -10$$

$$\underline{\quad -4 = -4}$$

$$-2x = -14$$

$$x = 7$$

Check (usually easy).

$5(7) + 4 = 7(7) - 10$;
$39 = 39$.

Always check in the original.

EXAMPLE 2—

The sum of two numbers is 10. The sum of twice one and triple the other is 27. Find them.

NOTE

If the sum of two numbers is 10; one is 6, the other would be $10 - 6$; if one were 9.5, the other would be $10 - 9.5$; if one is x, the other is $10 - x$.

SOLUTION—

One number is x; double this is $2x$ (double means multiply by 2). The other number is $10 - x$; triple that is $3(10 - x)$; triple means multiply by 3. Sum means add.

$2x + 3(10 - x) = 27 \qquad 2x + 30 - 3x = 27 \qquad -x = -3$

So $x = 3$. The other number is $10 - 3 = 7$.

EXAMPLE 3—

The difference between two numbers is 2. The sum is 25. Find them.

METHOD 1   One number $= x$. The second number is $x - 2$.

$x + x - 2 = 27 \qquad 2x = 29 \qquad x = 14.5 \qquad x - 2 = 12.5$

METHOD 2   One number is x. The second is $x + 2$.

$x + x + 2 = 27$. $2x + 2 = 27$. $2x = 25$.

$x = 12.5$; $x + 2 = 14.5$

LOTS OF NOTES

1. Notice whether you use x and $x + 2$ or x and $x - 2$, the difference is 2.

2. In one case you get the smaller one first; in one case you get the larger. However no matter which way, both give you the same answers.

3. The word "numbers" *doesn't necessarily mean integers.*

4. Fractional answers are OK also: 12.5 could have been 12½ or ²⁵⁄₂.

5. There are more ways, but not now.

This gives you a small sample of the problem with words. As we go on, we will do more. For now, let's do some graphing.

CHAPTER 5

A POINT WELL TAKEN: GRAPHING POINTS AND LINES, SLOPE, EQUATION OF A LINE

GRAPHING POINTS

We have the *x*-axis, the horizontal (left to right) axis, and the *y*-axis, vertical (up and down) axis.

The plural of axis is axes (long e). The picture is at the right.

To the right is the positive *x*-axis; to the left is the negative *x*-axis.

Up is the positive *y*-axis; down is the negative *y*-axis.

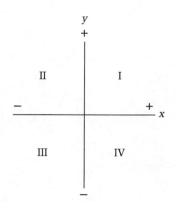

The graph is divided into four parts called *quadrants*. The upper right is the first quadrant since both *x* and *y* are positive. Counterclockwise says the second quadrant is the upper left, the third is the lower left, and the fourth is the lower right.

(*x*, *y*) is called an *ordered pair*. The *x* number is always given first, which is why it is called an ordered pair. *x* and *y* are called the *coordinates* of a point.

   x: abscissa or *first coordinate.*

   y: ordinate or *second coordinate.*

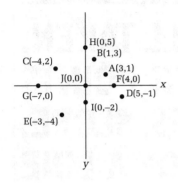

The point A(3,1), 3 to the right and 1 up, is not the same as point B(1,3), 1 to the right and 3 up. That is why it is called an ordered pair; order is important!!!!

C(−4,2) is 4 to the left and 2 up.

D(5,−1) is 5 to the right and 1 down.

E(−3,−4) is 3 to the left and 4 down.

F(4,0) is 4 to the right and 0 up.

G(−7,0) is 7 to the left and 0 down.

All points on the x axis have the y coordinate = 0.

H(0,5) is 0 to the left and 5 up.

I(0,−2) is 0 to the right and 2 down.

All points on the y axis have the x coordinate = 0.

J(0,0) is where the axes meet. It is called the origin.

GRAPHING LINES

We would like to graph lines. (The word line always means both infinite and straight.)

EXAMPLE I—

Graph $y = 2x + 3$.

Select three values for x. For each x value, find the y value. Graph each of the points. Connect the points. Label the line.

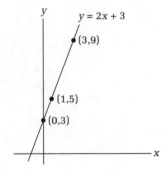

x	y	(x, y)
0	$2(0) + 3 = 3$	(0,3)
1	$2(1) + 3 = 5$	(1,5)
3	$2(3) + 3 = 9$	(3,9)

NOTES

1. The numbers can be but don't have to be consecutive.

2. Negative numbers are also OK.

3. Take numbers that are easy. Don't take, let us say, $x = 2.3456$.

4. Take numbers that are relatively close to the origin. You could let $x = 7000$, but the point will be on the moon.

EXAMPLE 2—

Graph $2x + 3y = 12$.   Step 1: Solve for y.

$$-2x \qquad = -2x$$

$$3y = -2x + 12$$

$$\frac{3y}{3} = \frac{-2}{3}x + \frac{12}{3} \text{ or}$$

$$y = \frac{-2}{3}x + 4$$

Pick values for x. In this case we will pick multiples of 3 so that we don't have any fractions for the point coordinates.

x	$y = \dfrac{-2}{3}x + 4$	(x, y)
0	$y = \dfrac{-2}{3}(0) + 4 = 4$	$(0,4)$
3	$y = \dfrac{-2}{3}(3) + 4 = 2$	$(3,2)$
6	$y = \dfrac{-2}{3}(6) + 4 = 0$	$(6,0)$

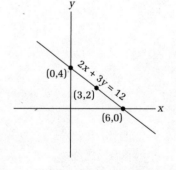

Notice the pattern on the chart; the ratio of the y's to the x's is the coefficient of x, $-\frac{2}{3}$. We will look at this soon.

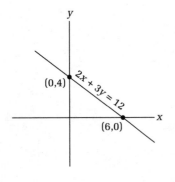

Example 2, $2x + 3y = 12$, revisited.

There is a short way to graph this line.

Any point on the x axis, the y coordinate is 0. This is called the x intercept.

Any point on y axis, $x = 0$. This is called the y intercept.

x intercept, $y = 0$. $2x + 3y = 12$. $2x + 3(0) = 12$. $x = 6$. We get the point $(6,0)$. (x coordinate always goes first.)

y intercept, $x = 0$. $2x + 3y = 12$. $2(0) + 3y = 12$. $y = 4$. $(0,4)$. Graph the two points and connect them.

This always works except in three cases.

EXAMPLE 3A—

Graph the points $(4,0)$, $(4,1)$, $(4,2)$, $(4,3)$. We have a vertical line. For every point on the line, the x coordinate is 4. The equation is $x = 4$.

All vertical lines are $x =$ something.

The y axis is $x = 0$. (Not what you might have thought.)

EXAMPLE 3B—

Graph the points $(-3,2)$, $(0,2)$ $(3,2)$, $(4,2)$.

We have a horizontal line. Each point has the y coordinate 2. The equation of the line is $y = 2$.

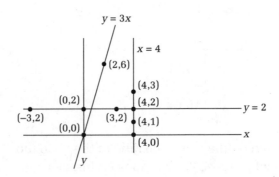

Horizontal lines have the equation $y = $ something.

The x axis is $y = 0$.

EXAMPLE 3C—

$y = 3x$. If we let $x = 0$, we get $y = 0$, and we get the point $(0,0)$. Let $x = $ anything, say 2. $y = 6$. We get $(2,6)$. Connect the two points.

In each case, we have only one intercept, which is both the x and y intercepts. All other lines have two distinct intercepts.

SLOPE

Before we find the equation of a line we wish to examine its SLOPE or slant.

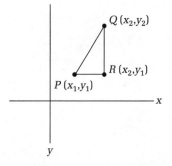

Given points $P(x_1, y_1)$ and $Q(x_2, y_2)$. Draw the line segment PQ, the horizontal line through P and the vertical line through Q meeting at R. Since Q and R are on the same vertical line segment, they have the same x value, x_2. Since P and R are on the same horizontal line, their y values are the same, y_1. The coordinates of R are $R(x_2,y_1)$. For Q and R, since the x values are the same, the length of QR is $y_2 - y_1$. If R is $(3,2)$ and Q is $(3,7)$, since the x values are the same, the length of QR is $7 - 2$. In general $y_2 - y_1$. Similarly the length of PR is $x_2 - x_1$.

Definition   SLOPE $= m = \dfrac{\text{change in } y}{\text{change in } x} = \dfrac{\Delta y}{\Delta x} = \dfrac{y_2 - y_1}{x_2 - x_1}$.

Δ: **delta (capital), a Greek letter.**

EXAMPLE 1—

Find the slope of the line joining $(1,3)$ and $(4,7)$.

Let $(x_1,y_1) = (1,3)$ and $(x_2,y_2) = (4,7)$.

$$m = \frac{y_2 - y_1}{x_2 - x_1} = \frac{7 - 3}{4 - 1} = \frac{4}{3}$$

The 1 and the 2 of x_1 or y_2 are called *subscripts*.

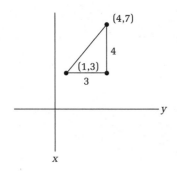

NOTES

1. As you go left to right, if you go up the line it has a positive slope.

2. You should keep writing the letters with subscripts until you can find the slope without writing the formulas.

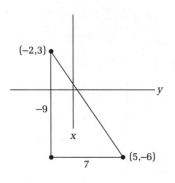

EXAMPLE 2—

Find the slope of the line joining $(-2,3)$ and $(5,-6)$.

Let $(x_1, y_1) = (-2,3)$ and $(x_2, y_2) = (5,-6)$.

$$m = \frac{y_2 - y_1}{x_2 - x_1} = \frac{-6 - 3}{5 - (-2)} = -\frac{9}{7}$$

NOTES

1. As you go left to right, if you go down the line, it has a negative slope.

2. Be careful of the minus signs!

EXAMPLE 3—

Find the slope of the line through $(-3,2)$ and $(5,2)$.

$$m = \frac{2 - 2}{5 - (-3)} = \frac{0}{8} = 0$$

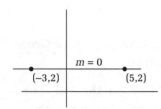

NOTES

1. Horizontal lines have *slope m = 0*.

2. The *equation* of this horizontal line is $y = 2$.

EXAMPLE 4

Find the slope of the line joining (3,2) and (3,7).

$$m = \frac{7-2}{3-3} = \frac{5}{0}, \text{ undefined}$$

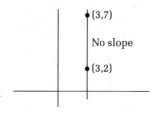

NOTES (TWO AS USUAL)

1. Vertical lines have *no slope* (sometimes called *infinite* slope).

2. The *equation* of this vertical line is $x = 3$.

EQUATION OF A LINE

Although there is not too much algebra in this part, I believe many will consider this topic kind of difficult. (If you don't, don't worry. Congratulate yourself.)

We have already graphed a line in *standard form, Ax + By = C*, with A and B both not 0. However two other forms are more useful. Let us show how we get them, and then use both.

Point-slope form: Given slope m and point (x_1, y_1) as pictured at the right. Let (x, y) represent any other point on the line. By definition of the slope,

$$m = \frac{\text{change in } y}{\text{change in } x} = \frac{y - y_1}{x - x_1}$$

That's it!

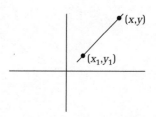

EXAMPLE 1

Find the equation of a line through point (5,–7) with slope ⅔.

$$m = \frac{y - y_1}{x - x_1} \qquad \frac{2}{3} = \frac{y - (-7)}{x - 3} \qquad \text{or} \qquad \frac{2}{3} = \frac{y + 7}{x - 3}$$

I allow my students to leave their answers like this, but your teacher may want you to write the answer in standard form.

One more form is very important. It is called *slope intercept* form. We will get it from point-slope form.

$$\frac{m}{1} = \frac{y - y_1}{x - x_1}$$ Cross-multiply.

$y - y_1 = m(x - x_1)$ (In some books, this is their form of point-slope; I don't like it as much.)

$y - y_1 = mx - mx_1$ Distribute and add y_1 to both sides, we get . . .

$y - y_1 = mx + y_1 - mx_1$

Now m and x_1 are numbers; their product mx_1 is a number. y_1 is also a number; a number minus a number, $y_1 - mx_1$, is a number. We will give it a new name, b!

$y = mx + b$

If we solve for y, the coefficient of x, which is m, is the slope and b is the y intercept. (If $x = 0$, $y = b$.)

EXAMPLE 2—

Find the equation of the line if the slope is $-\frac{2}{3}$ and the y intercept is $\frac{34}{19}$.

$y = mx + b$ $7 = (-\frac{2}{3})x + (\frac{34}{19})$

However point slope is easier about 90% of the time. I hope the next example will convince you.

EXAMPLE 3—

Given points $(3,5)$ and $(10,16)$. Find the equation of the line, and write the answer in standard form.

In any case we have to find the slope $m = (16 - 5)/(10 - 3) = \frac{11}{7}$.

METHOD I

$$m = \frac{y - y_1}{x - x_1}.$$

$$\frac{11}{7} = \frac{y - 5}{x - 3}$$

Although both points work, always use the one that gives the least amount of work.

$11(x - 3) = 7(y - 5)$ Cross-multiply.

$11x - 33 = 7y - 35$

$11x - 7y = -2$

METHOD 2

$$y = mx + b.$$

$$(5) = \frac{11}{7}(3) + b$$

$$5 = \frac{33}{7} + b$$

$$\frac{35}{7} = \frac{33}{7} + b \ (5 = {}^{35}\!/_{7})$$

$$b = \frac{35}{7} - \frac{33}{7} = \frac{2}{7}$$

$$y = \frac{11}{7}x + \frac{2}{7}$$

$$7y = 11x + 2$$

$-11x + 7y = 2$ Multiply both sides by -1 since standard form is written with the first coefficient positive.

$11x - 7y = -2$ Whew!!!!

Method 1 is much easier because there are fewer fractions to worry about. Let's do one more.

EXAMPLE 4—

Find the equation of the line with x intercept -5 and y intercept 9. x intercept -5 means the point $(-5,0)$. y intercept 9, means the point $(0,9)$. $m = (9 - 0)/(0 - (-5)) = \frac{9}{5}$.

The equation is $\dfrac{9}{5} = \dfrac{y - 9}{x - 0}$ or $\dfrac{9}{5} = \dfrac{y - 0}{x + 5}$ or $y = \dfrac{9}{5}x + 9$ or $9x - 5y = -9$.

More? See *Algebra for the Clueless*. Next, percentages.

CHAPTER 6

RATIOS, PROPORTIONS, AND PERCENTAGES

RATIOS AND PROPORTIONS

A *ratio* is the comparison of two things.

EXAMPLE 1—

Write the ratio of 2 to 7: either 2/7 or 2:7.

EXAMPLE 2—

Write the ratio of 2 feet to 11 inches. 2 feet = 24 inches
(12 inches = 1 foot): 24/11 or 24:11

EXAMPLE 3—

Write the ratio of 3 cats to 4 dogs. I don't know why
you would want to do this, but the answer is 3/4.

Proportion   Two ratios equal to each other $a/b = c/d$.

A proportion $a/b = c/d$ is true if $ad = bc$. (This is cross-
multiplication!!!!!)

EXAMPLE 4—

Solve for x: $4/x = 5/7$. $5x = 28$. $x = 28/5$.

We've done this before. We just gave the problem a name.

PERCENTAGES

We are going to do percentages a little different. But first we need to take care of the basics.

CHANGING

Percent means $\frac{1}{100}$ of the whole. 100% = 1 (1 whole thing).

% to fraction to decimal: Move the decimal points two places to the left and drop the percent sign. This is true since 43% means $\frac{43\%}{100\%} = \frac{43}{100} = .43$!!! Here are two more.

EXAMPLE 1

$3\% = 3.\% = \frac{3\%}{100\%} = \frac{3}{100} = .03$; $.23\% = \frac{.23\%}{100\%} = \frac{.23}{100} = \frac{23}{10,000} = .0023$.

Decimal to fraction to %: Move the decimal point two places to the right and add the % sign.

EXAMPLE 2

$.543 = \frac{543}{1000} = \frac{54.3}{100} = 54.3\%$; $4.39 = 4\frac{39}{100} = 439/100 = 439\%$; $32.9 = 32\frac{9}{10} = \frac{329}{10} = \frac{3290}{100} = 3290\%$; $23 = 23.00 = 2300\%$.

If you want a fraction for an answer, the fraction should be reduced.

EXAMPLE 3

Change 46% to a fraction: $46\% = \frac{46\%}{100\%} = \frac{46}{100} = \frac{23}{50}$.

EXAMPLE 4

Change .24 to a fraction: $.24 = \frac{24}{100} = \frac{6}{25}$.

OK let's get to some percentage problems

PERCENTAGE PROBLEMS

If you have had percentage problems in the past, here's a method you might like. Almost all of the students who had problems with percentages really like this method.

Draw a pyramid and label it as shown.

That's all there is to it.

Let's do the three basic problems.

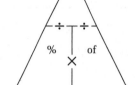

EXAMPLE 1—

What is 14% of 28?

14% = .14 which we put in the % box.

28 goes in the "of" box.

The pyramid says multiply .14 × 28 = 3.92. That's all!!

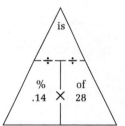

EXAMPLE 2—

12% of what is 1.8?

12% = .12 in the percent box.

1.8 in the "is" box. The pyramid says divide 1.8 by .12.

$.12\overline{)1.8} = 12\overline{)180} = 15$

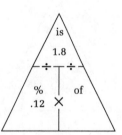

EXAMPLE 3—

9 is what percent of 8?

9 goes in the "is" box. 8 goes in the "of" box.

The pyramid says 9 divided by 8 (and then change to a percent).

$8\overline{)9.0000} = 1.125 = 112.5\%$

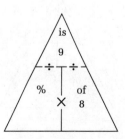

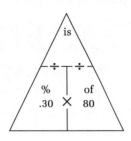

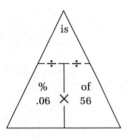

EXAMPLE 4

An $80 radio is discounted 30%. How much do you pay?

We must take 30% of $80 and then subtract from $80.

According to the pyramid, the discount is .30 × $80 = $24.

Then $80 − $24 = $56 is the cost to you.

Suppose your state has a 6% sales tax.

We take 6% of $56 and add it to the cost.

.06 × $56 = $3.36. The total cost is $56.00 + $3.36 = $59.36

NOTE

If you have a 30% discount, if we took 70% of $80 (70% = 100% − 30%), we would have immediately gotten the cost.

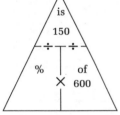

EXAMPLE 5

A TV regularly sells for $600. If it sells for $450, what is the percent discount?

$600 − $450 = $150 is the discount

$^{150}/_{600} = \frac{1}{4} = 25\%$ is the percent discount

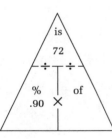

EXAMPLE 6

$72 is 90% of the original cost. What was the original cost?

$^{72}/_{.90} = .9\overline{)72} = 9\overline{)720} = \$80$

I hope this gives you a different look at percentages. We use it a lot in our world. We'll probably do some more percentages later.

Now algebra and arithmetic are getting a little tiring. Let's talk a little about geometry for a while.

CHAPTER 7

SOME BASICS ABOUT GEOMETRY AND SOME GEOMETRIC PROBLEMS WITH WORDS

Let's explain a few basic terms, how to label them and picture them. Terms like point and line are actually undefined terms. (They are terms we understand, but if you try to define them you will fail.)

A point is indicated by one capital letter.

A line is indicated by $\overleftrightarrow{AB}$.

A ray is all points on a line on one side of a line.

Ray $\overrightarrow{AB}$ has A as the endpoint.

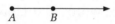

$\overrightarrow{BA}$ has the endpoint.

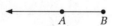

Line segment: all points on a line between two points on that line.

Two notations: $\overline{AB}$ or AB; I'll use AB; it's simpler!

Angle: two rays with a common endpoint (called the *vertex*).

B is the vertex. $\angle$ is the symbol for angle. This angle is read $\angle ABC$ or $\angle CBA$ or $\angle B$. (The vertex is always in the middle.)

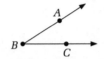

We cannot always use only one letter for an angle. In the picture below there are *three* angles with B as its vertex: $\angle ABD$, $\angle DBC$, and $\angle ABC$!

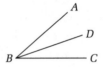

Technically we cannot say two angles are equal. How silly!! Either angles are congruent, equal in every way, written $\angle C \cong \angle D$ or these are equal in degrees (m = measure), written $m\angle C = m\angle D$. I will just say $\angle C = \angle D$ and hope by the time you take geometry, that's how it will be.

Sorry, but this is how a course in geometry starts. We still need some more words and more facts.

Once around a circle is 360 degrees, written 360°.

Half way around a circle is 180 degrees.

Two lines that are perpendicular form one or more right angles, 90° angles, each is ¼ around a circle.

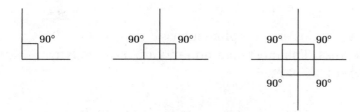

An angle of less than 90 degrees is called *acute*.

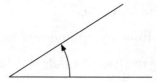

An angle more than 90 degrees but less than 180 degrees is called *obtuse*.

A 180 angle is called a *straight angle*.

An angle that is more than 180 degrees but less than 360 is called a *reflex angle*.

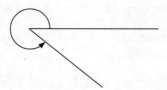

Two angles are supplementary if they add to 180 degrees.

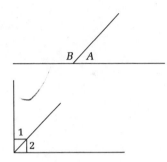

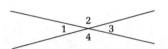

The most common picture is at the left.

$\angle A + \angle B = 180°$.

Two angles are complementary if they add to 90 degrees. The most common picture is at the left. $\angle 1 + \angle 2 = 90°$.

The sum of the angles of a triangle is 180 degrees.

Vertical angles are equal. 1 and 3 are equal. So are 2 and 4. Look for the letter "X."

Parallel lines are lines in a plane that never meet.

A line intersecting the two parallels is called a transversal.

If two parallel lines are cut by a transversal, alternate interior angles are equal. On this picture we have angles 3 and 6, 4 and 5. $\angle 3 = \angle 6$, $\angle 4 = \angle 5$. Look for the letter "Z" or backward "Z."

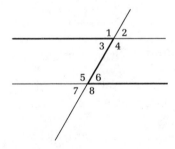

If two parallel lines are cut by a transversal, alternate exterior angles are equal. $\angle 1 = \angle 8$, $\angle 2 = \angle 7$.

If two parallel lines are cut by a transversal, corresponding angles are equal.

The following pairs are equal: 1 and 5 (upper left), 2 and 6 (upper right), 4 and 8 (lower right), 3 and 7 (lower left).

Look for the letter "F" (could be upside down, backward, etc.). The F for angles 4 and 8 are pictured.

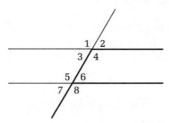

If two parallel lines are cut by a transversal, interior angles on the same side of the transversal are supplementary. $\angle 4 + \angle 6 = 180°$ and $\angle 3 + \angle 5 = 180°$.

Look for the letter "C" or backward "C" (a square "C").

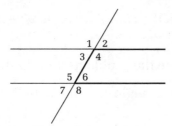

Enough already! Let's do some problems. Most geometric problems are pretty easy.

EXAMPLE 1—

The angles of a triangle are in the ratio of 3 to 4 to 5. Find the angles.

Sometimes we write this ratio 3:4:5. If one angle is $3x$, the second is $4x$, and the third is $5x$. The equation is $3x + 4x + 5x = 180$. $12x = 180$. $x = 15$ degrees.

Angle 1 is $3(15) = 45$ degrees. Angle 2 is $4(15) = 60$ degrees. Angle 3 is $5(15) = 75$ degrees.

EXAMPLE 2—

One angle is 6 more than twice another. If they are supplementary, find the angles.

Let x = one angle; the second angle = $2x + 6$. Supplementary means they add to 180 degrees. $x + 2x + 6 = 180$. $3x + 6 = 180$. $3x = 174$. $x = 58$.

One angle is 58 degrees. The other is $2(58) + 6 = 122$ degrees. Notice $58 + 122 = 180$.

EXAMPLE 3—

The supplement of an angle is 6 times the complement. Find the angle.

We need a little preparation here.

If 11 is the angle, the complement is $90 - 11$, and the supplement is $180 - 11$.

If x is the angle, the complement is $90 - x$, and the supplement is $180 - x$.

Now let's rewrite the problems and do it.

Supplement of an angle is 6 times the complement (of the angle).

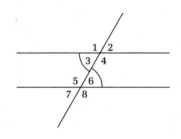

$180 - x = 6(90 - x)$

$180 - x = 540 - 6x$. $5x = 360$. $x = 72$.

Notice its supplement, $180 - 72 = 108$ degrees, is 6 times the complement $90 - 72 = 18$ degrees.

EXAMPLE 4—

With parallel lines angle 3 is $10x + 2$ and angle 6 is $2x + 26$. Find all the angles. We find the letter Z, alternate interior angles.

So $10x + 2 = 2x + 26$. $8x = 24$. $x = 3$.

$10(3) + 2 = 32$ degrees. Of course $2(3) + 26 = 32$ degrees.

In fact, with these word problems, you don't actually need to know the names. You just have to see that angles that look equal are equal. If they are not equal they are supplementary!!!!! $\angle 2, \angle 3, \angle 6, \angle 7 = 32°$.

So, $\angle 1, \angle 4, \angle 5, \angle 8 = 180° - 32° = 142°$.

EXAMPLE 5—

Solve for x.

Since these are vertical angles, we get $12x = 2x + 200$.

$x = 20$ degrees. Each of the marked vertical angles is 240 degrees: $2(20) + 200 = 240$.

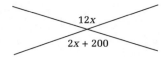

EXAMPLE 6—

Finally, let's look at one tricky one. Find w, y, z.

It looks like the above problem. However . . . we get $5w = 3y + 5$. We can't do that. $2y = w + z$. We can't do that either. We have to see on one side of a straight line $2y$ and $3y + 5$ add to 180 degrees!!!

$2y + 3y + 5 = 180$. $5y + 5 = 180$. $5y = 175$. $y = 35$ degrees.

$3y + 5 = 3(35) + 5 = 110 = 5w$. $w = 22$ degrees.

$2y = 2(35) = 70 = w + z = 22 + z$. $z + 22 = 70$. $z = 48$.

Problems like this can be found on the SAT (this is a harder one).

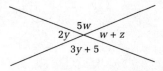

Let's learn a little about three-sided and four-sided figures (triangles and quadrilaterals).

CHAPTER 8

TRIANGLES, SQUARE ROOTS, AND GOOD OLD PYTHAGORAS

We are starting to study polygons. *Polygons* are closed figures in the plane with sides that are line segments.

Let's first go over triangles.

Triangle: A polygon with three sides. Triangles also have three angles. We have already stated that the sum of the angles of a triangle is 180 degrees. Later on in this chapter, we will give a sample proof from geometry.

Triangles can be classified two ways: by sides and by angles.

A *scalene* triangle is a triangle that has three **unequal** sides. It also has three unequal angles.

The ever popular *isosceles* triangle is a triangle with two equal sides.

We have triangle *ABC,* △ *ABC,* at the side. *AB* and *BC* are the legs. They are equal.

∠*A* and ∠*C* are called **base angles.** They are also equal.

The **base** is *AC.* It can be equal to a leg, but it usually is larger or smaller. ∠*B* is called the vertex angle. It can equal a base angle, bigger than a base angle if the base is bigger than a leg, or is smaller than a base angle, if the base is smaller than a leg.

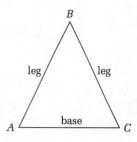

Last, we have an equilateral triangle, with all sides equal.

All angles are equal. They each equal 60°.

Triangles can also be classified according to angles.

An *acute* triangle is a triangle with each angle less than 90°.

The most used triangle, a *right* triangle, is a triangle with one right (90°) angle.

The side *AB* opposite the right angle *C* is called the **hypotenuse.**

The sides, *BC* and *AC,* opposite the acute angles are called **legs.**

They may or may not be equal.

An *obtuse* triangle is a stupid triangle. Just a joke.

It is a triangle with one obtuse angle, an angle between 90° and 180°.

Triangles can be a combination of the two.

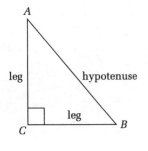

EXAMPLE I

Give one possible set of angles (there are many) of a triangle that are the following:

   A. Acute and scalene

   B. Right and scalene

   C. Obtuse and scalene

   D. Acute and isosceles

   E. Right and isosceles

   F. Obtuse and isosceles

EXAMPLE I SOLUTIONS—

   A. Angles of 45°, 55°, 80°: angles all less than 90°, so the triangle is acute; all angles are unequal which means the sides are unequal which means it is scalene.

B. 30°, 60°, 90°: one right angle with all sides unequal; this is a very important triangle which we will look at a little in this book.

C. 25°, 38°, 117°: obtuse (one angle more than 90°) and all sides unequal.

D. 58°, 58°, 64°: all angles less than 90° (acute); two angles equal (isosceles) since this means two sides equal.

E. 45°, 45°, 90°: the only isosceles right triangle; also very important; we will deal with this again later.

F. 16°, 16°, 148°: one obtuse angle; the other two acute angles equal.

NOTE

There is only one equilateral, acute equilateral, since all the angles must be 60 degrees.

EXAMPLE 2—

In an isosceles triangle, the vertex angle is 6 degrees more than the base angle. Find all the angles. Since the base angles are equal, both equal x; the vertex angle is $x + 6$.

The equation is $x + x + x + 6 = 180$. $3x + 6 = 180$. $3x = 174$. $x = 58$.

The angles are 58°, 58°, 58° + 6° = 64°.

Most geometry (and trig) word problems are very easy. We will do more problems on triangles when we get to later chapters. In regular geometry, many weeks are spent on triangles. We will do more on triangles later, but not nearly as much as you would find in geometry.

Although this would not be part of most prealgebra or pregeometry courses, let's give one proof that would be part of geometry. Let's show a fact we have already been given.

Theorem: a provable law. Postulate: a law taken to be true without proof.

Theorem   The sum of the angles of a triangle is 180°.

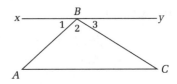

Statement	Reason
1. Draw line $\overleftrightarrow{XY}$ through B parallel to AC.	1. Postulate: Through a point not on a line (segment) one and only one line can be drawn parallel to another line (segment).
2. $\angle 1 + \angle 2 + \angle 3 = 180°$.	2. The sum of the angles on one side of a line $= 180°$.
3. $\angle 1 = \angle A$, $\angle 3 = \angle C$.	3. If two lines are parallel, alternate interior angles are equal. (See the Z and the backward Z?!!!!)
4. $\angle A + \angle 2 + \angle C = 180°$. (Look at the picture; this says the sum of the angles of a triangle is 180 degrees!!!)	4. Substitute $\angle A$ for $\angle 1$ and $\angle C$ for $\angle 3$ in step 2.

Believe it or not, the way you think through a proof is the way you answer an English essay or a history essay. Only if you become a major in math or math-related areas will you do a lot of proofs. (Now please don't laugh too loud.) Maybe, with the help of my books, you will become so good at math you will want to do things with it. Math is incredibly wonderful, but to get to the point where you can see how great it is, you must learn your basics well!!!!!).

Let's do some work with square roots and right triangles.

Square root, $\sqrt{}$, is perhaps the best liked symbol in all math. Let's define it.

Definition of *square root:* $\sqrt{a} = b$ if $(b) \times (b) = a$. $\sqrt{9} = 3$ since $(3) \times (3) = 9$.

EXAMPLE 1—

A. $\sqrt{16} = 4$; B. $-\sqrt{16} = -4$; C. $\sqrt{-16}$ has no real answer.

  A. True since $(4)(4) = 16$.

  B. True since $-(4)(4) = -16$.

  C. $(4)(4) = 16$ and $(-4)(-4) = 16$. There is no real number b such that $b^2 = -16$.

NOTE

It would be nice to know all the following because most come up over and over and over, but I would not sit down to memorize them; just keep the list in front of you. You will be surprised how many you will soon know.

$\sqrt{1} = 1$, $\sqrt{4} = 2$, $\sqrt{9} = 3$, $\sqrt{16} = 4$, $\sqrt{25} = 5$, $\sqrt{36} = 6$, $\sqrt{49} = 7$, $\sqrt{64} = 8$

$\sqrt{81} = 9$, $\sqrt{100} = 10$, $\sqrt{121} = 11$, $\sqrt{144} = 12$, $\sqrt{169} = 13$, $\sqrt{196} = 14$, $\sqrt{225} = 15$, $\sqrt{256} = 16$

$\sqrt{289} = 17$, $\sqrt{324} = 18$, $\sqrt{361} = 19$, $\sqrt{400} = 20$, $\sqrt{441} = 21$, $\sqrt{484} = 22$, $\sqrt{529} = 23$, $\sqrt{576} = 24$

$\sqrt{625} = 25$, $\sqrt{676} = 26$, $\sqrt{729} = 27$, $\sqrt{784} = 28$, $\sqrt{841} = 29$, $\sqrt{900} = 30$, $\sqrt{961} = 31$, $\sqrt{1024} = 32$

We would like to simplify square roots.

EXAMPLE 2—

Simplify $\sqrt{12}$. $\sqrt{12} = \sqrt{(2)(2)(3)} = 2\sqrt{3}$.

Reason? Two primes on the inside become one on the outside. Why? $\sqrt{(2)(2)}(= \sqrt{4}) = 2$.

EXAMPLE 3—

Simplify A. $\sqrt{125}$; B. $\sqrt{72}$; C. $6\sqrt{98}$.

A. $\sqrt{125} = \sqrt{(5)(5)(5)} = 5\sqrt{5}$; two come out as one; the third five stays.

B. $\sqrt{72} = \sqrt{(2)(2)(2)(3)(3)} = (2)(3)\sqrt{2} = 6\sqrt{2}$; one 2 and one 3 come out and are multiplied; one 2 stays under the square root sign.

C. $6\sqrt{98} = 6\sqrt{2(7)(7)} = 6(7)\sqrt{2} = 42\sqrt{2}$.

There are two square roots that we want to know the approximate value, because they come up so much!!! $\sqrt{2} = 1.414$ and $\sqrt{3} = 1.732$, the year George Washington was born. My Aunt Betty (one of two) once lived at 1732. I could never forget that house number!!!

We wish to combine terms; it is the same as adding (subtracting) like terms!!!!!

EXAMPLE 4—

Combine $4\sqrt{5} + 7\sqrt{3} + 2\sqrt{5} - 4\sqrt{3}$.

$4\sqrt{5} + 2\sqrt{5} + 7\sqrt{3} - 4\sqrt{3} = 6\sqrt{5} + 3\sqrt{3}$

You cannot simplify any more without a calculator.

EXAMPLE 5—

Simplify and combine $2\sqrt{27} + 5\sqrt{18} + 3\sqrt{32}$.

$2\sqrt{3(3)(3)} + 5\sqrt{2(3)(3)} + 3\sqrt{2(2)(2)(2)(2)} = 6\sqrt{3} + 15\sqrt{2} + 12\sqrt{2} = 6\sqrt{3} + 27\sqrt{2}$

Only like radicals can be combined.

Let's do a little multiplication: $a\sqrt{b} \times c\sqrt{d} = ac\sqrt{bd}$ if b and d are not negative.

EXAMPLE 6—

A. $3\sqrt{5} \times 4\sqrt{7}$; B. $7\sqrt{6} \times 10\sqrt{8}$.

A. $12\sqrt{35}$

B. $70\sqrt{3(2)(2)(2)(2)} = 70(2)(2)\sqrt{3} = 280\sqrt{3}$. It is silly to multiply out first, because you then have to break it down; break it down at once!!!!

A little division.

EXAMPLE 7—

$$\sqrt{\frac{4}{9}} = \frac{\sqrt{4}}{\sqrt{9}} = \frac{2}{3}.$$

A little rationalization of the denominator (no square roots on the bottom).

EXAMPLE 8—

Rationalize the denominator   A. $\dfrac{5}{\sqrt{7}}$; B. $\dfrac{6}{\sqrt{32}}$.

A. $\dfrac{5}{\sqrt{7}} \times \dfrac{\sqrt{7}}{\sqrt{7}} = \dfrac{5\sqrt{7}}{7}$.

B. Simplify the bottom first

$$\frac{6}{\sqrt{32}} = \frac{6}{\sqrt{2(2)(2)(2)(2)}} = \frac{6}{4\sqrt{2}} = \frac{3}{2\sqrt{2}} = \frac{3}{2\sqrt{2}} \cdot \frac{\sqrt{2}}{\sqrt{2}} = \frac{3\sqrt{2}}{4}.$$

Let's do some problems about the most famous theorem in all of math, the Pythagorean theorem (named after the Greek mathematician, Pythagoras). I'm so excited!!!

Pythagorean Theorem   In a **right** triangle, $c^2 = a^2 + b^2$. (Hypotenuse)^2 = (leg)^2 + (other leg)^2.

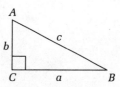

EXAMPLE 1—

Find x.

$x^2 = 4^2 + 6^2$. $x^2 = 52$. $x = \sqrt{52} = \sqrt{2(2)(13)} = 2\sqrt{13}$.

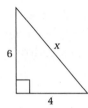

EXAMPLE 2—

Find x.

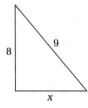

$9^2 = x^2 + 8^2$ or $8^2 + x^2$ The hypotenuse (squared) is always by itself since it is the largest side.

$x^2 = 9^2 - 8^2 = 81 - 64 = 17$

$x = \sqrt{17}$.

That's really all it is to the basics. However in geometry and trig you'll need some more. The following are called Pythagorean triples. The side listed last is the hypotenuse.

3–4–5 group ($3^2 + 4^2 = 5^2$): 3,4,5 . . . 6,8,10, . . . 9,12,15, . . . 12,16,20, . . . 15,20,25.

5–12–12 group: 5,12,13, . . . 10,24,26.

Also 8,15,17 and 7,24,25.

Are there any more? Yes, an infinite number. Some are 9,40,41 . . . 11,60,61 . . . 20,21,29. Why should you memorize only the nine? Because the first nine I listed

come up over and over and over again. The other three
I listed have come up once every 7 or 8 years. We cannot memorize everything. We have to memorize the
most important ones.

EXAMPLE 3—

The legs are 3 and 5. Find the hypotenuse.
  This is to trick you. The missing side is **not** 4.

$$x^2 = 3^2 + 5^2 \qquad x^2 = 9 + 25 \qquad x = \sqrt{34}!$$

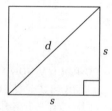

  There are two special right triangles you will need,
probably later.
  45°–45°–90° triangle. It comes from a square.

$$d^2 = s^2 + s^2 \qquad d^2 = 2s^2 \qquad \sqrt{d^2} = \sqrt{2s^2} \qquad d = s\sqrt{2}$$

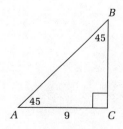

In a 45°–45°–90° triangle:

1. The legs are equal.

2. Leg to hypotenuse? $\times \sqrt{2}$.

3. Hypotenuse to leg? $\div \sqrt{2}$.

EXAMPLE 1—

Given $AC = 9$.

  $AB = 9$, since both legs are equal.

  $AB = 9\sqrt{2}$, the leg $\times \sqrt{2}$.

EXAMPLE 2—

Let $AB = 10$.

$$AC = BC = \frac{10}{\sqrt{2}} = \frac{10}{\sqrt{2}} \times \frac{\sqrt{2}}{\sqrt{2}} = \frac{10\sqrt{2}}{2} = 5\sqrt{2}$$

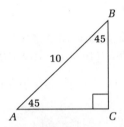

Let's do the 30°–60°–90° triangle.

Draw equilateral triangle ABC base AB.

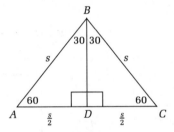

Draw $BD \perp AC$. In geometry you will show that both triangles are the same. Let $AB = BC = AC = s$ (in feet or inches or meters). s = side, h = height. Since $AD = DC$, both $= s/2$.

In right triangle ABD, $AB^2 = AD^2 + BD^2$ or $s^2 = (1/2s)^2 + h^2$.

$$h^2 = 1s^2 - \frac{1}{4}s^2 \qquad \sqrt{h^2} = \sqrt{\frac{3}{4}s^2} \qquad h = \frac{\sqrt{s^2}\sqrt{3}}{\sqrt{4}} = \frac{1}{2}s\sqrt{3}$$

Also since $BD \perp AC$, $\angle BDA$ and $\angle BDC = 90°$.

Since the triangle was equilateral, $\angle A = \angle C = 60°$.

So the angles $\angle ABD$ and $\angle CBD$ must $= 30°$.

Let us translate this into English:

Opposite the 90°, the hypotenuse is *s*.

Opposite the 30° is *s*/2 (short leg).

Opposite the 60° = (*s*/2)$\sqrt{3}$ (long leg).

How to find the missing sides:

1. Always get the short leg first, if it is not given.

2. Short to hypotenuse? × 2.

3. Hypotenuse to short? ÷ 2.

4. Short to long? ×$\sqrt{3}$.

5. Long to short? ÷$\sqrt{3}$.

EXAMPLE 1—

Short = 9.

Hypotenuse = 2 × 9 = 18.

Long = 9 × $\sqrt{3}$ = 9$\sqrt{3}$.

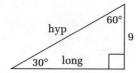

EXAMPLE 2—

Hypotenuse = 17.

Find the short first: hypotenuse/2 = 17/2 or 8.5

Long 8.5 × $\sqrt{3}$ = 8.5$\sqrt{3}$.

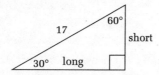

EXAMPLE 3—

Long = 12.

$$\text{Short} = \frac{\text{long}}{\sqrt{3}} = \frac{12}{\sqrt{3}} \times \frac{\sqrt{3}}{\sqrt{3}} = 4\sqrt{3}.$$

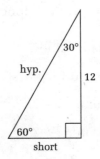

Hypotenuse = short $\times$ 2 = $8\sqrt{3}$.

That's all for now. Let's look at four-sided figures for a while.

CHAPTER 9

RECTANGLES, SQUARES, AND OUR OTHER FOUR-SIDED FRIENDS

We are about to talk about quadrilaterals. I am not sure who needs this chapter, but everyone needs these facts at some time. If you don't need this chapter right now, skip to the next chapter.

QUADRILATERAL   A four-sided polygon.

The sum of the angles of a quadrilateral is always 360°.

All quadrilaterals have two **diagonals,** a line drawn from one vertex to the opposite vertex. In the figure on the side, *AC* and *BD* are diagonals.

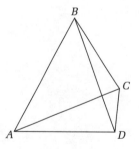

There are many kinds of quadrilaterals. One is a parallelogram.

Parallelogram: a quadrilateral with the opposite sides parallel. The following are important properties of a parallelogram, all of which are proven in a good geometry course:

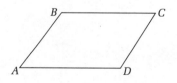

1. The opposite sides are equal: *AB* = *CD* and *AD* = *BC.*

2. Opposite angles are equal: ∠*A* = ∠*C*, ∠*B* = ∠*D*.

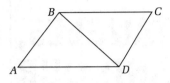

Bisect: divide into two equal parts.

3. Consecutive angles are supplementary:

$\angle A + \angle B = 180°$ $\angle B + \angle C = 180°$
$\angle C + \angle D = 180°$ $\angle D + \angle A = 180°$

4. One diagonal divides the parallelogram into two congruent (identical in every way) triangles.

$\triangle ABD \cong \triangle CDB$

5. The diagonals bisect each other:

$AE = EC$, and $BE = ED$

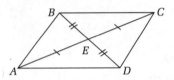

Let's get to one we know better.

Rectangle: A parallelogram with all right angles. A rectangle has properties 1 to 5 plus 6.

6. The diagonals are equal: e.g. $AC = BD$

Rhombus: A parallelogram with equal sides (a squashed square). A rhombus has properties 1 to 5, not 6, but 7.

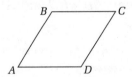

7. The diagonals are perpendicular: e.g. $AC \perp BD$

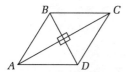

Square: A rectangle with equal sides. A square has all seven of the above properties. When I wrote the first time, I forgot the square, the easiest of the figures. I can't imagine why?! We will do a lot about squares and rectangles in the next chapter and probably after that.

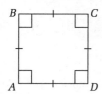

Trapezoid: A quadrilateral with exactly one pair of parallel sides. e.g. $AD \parallel BC$

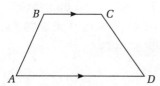

Isosceles trapezoid: A trapezoid with nonparallel sides equal.

$AD \parallel BC$ and $AB = CD$

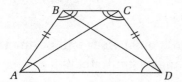

1. The diagonals are equal: $AC = BD$.

2. Base angles are equal: $\angle BAD = \angle CDA$ and $\angle ABC = \angle BCD$ (there are two bases).

EXAMPLE 1—

Let *ABCD* be a rectangle.

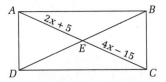

Let $AE = 2x + 5$. Let $EC = 4x - 15$. Find *BD*.

In a parallelogram (a rectangle is a special kind of parallelogram), the diagonals bisect each other. $2x + 5 = 4x - 15$. $-2x = -20$. $x = 10$. $AE = 2(10) + 5 = 25$. $EC = 4(10) - 15 = 25$. Diagonal $AC = AE + EC = 25 + 25 = 50$.

In a rectangle both diagonals are equal. $BD = AC = 50$.

EXAMPLE 2—

In an isosceles trapezoid, the bottom base angle is $x + 6$; the other bottom base angle is $2x - 17$. Find all the angles. In an isosceles triangle base angles are $x + 6 = 2x - 17$. $x = 23$. So each bottom base angle is 29° (23 + 6). Since the bases are parallel, each top base angle is the supplement since interior angles on the same side of the transversal are supplementary.

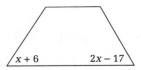

Each top angle is $180° - 29° = 151°$. Notice 29° + 29° + 151° + 151° = 360°. The sum of the angles of all quadrilateral must always be 360°.

Let's do more familiar problems: areas and perimeters of triangle, squares, rectangles, and others.

CHAPTER 10

SECURING THE PERIMETER AND AREAL SEARCH OF TRIANGLES AND QUADRILATERALS

We wish to find the area, the amount inside, and the perimeter, the distance around. The perimeter is measured in linear measurements such as inches, feet, yards, and miles (mi) and millimeters, centimeters, meters, and kilometers.

The area of a rectangle is a postulate (a law taken to be true without proof).

The formula for the area of a rectangle, $A = $ base times height or length $\times$ width. In symbols $A = b\,h$ or $l\,w$.

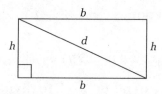

Perimeter: add all the sides.

$$P = b + h + b + h = 2b + 2h \qquad \text{or} \qquad 2(b + h)$$

12 inches (in) = 1 foot (ft)
3 ft = 1 yard (yd)
5280 ft = 1760 yd = 1 mile

Metric:

milli means 1/1000

centi means 1/100

kilo means 1000

10 millimeters mm = 1 centimeter (cm)

100 cm = 1 meter (m)

1000 m = 1 kilometer (km)

To go from larger to smaller, multiply.

To go from smaller to larger, divide, for example, 7 ft = 7 × 12 = 84 in

27 ft = 27/3 = 9 yd

2700 mm = 2700/1000 = 2.7 m

3.45 km = 3.45(1000) = 3450 m

Also:

2.54 cm ≈ 1 in; 1 cm ≈ .4 in

39.37 in ≈ 1 m; 8 km ≈ 5 mi.

≈ approximately equal.

EXAMPLE 1—

A rectangle has a base of 8 meters and a height of 5 meters. Find the perimeter, area, and the diagonal.

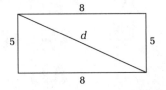

$p = 2b + 2h = 2(8) + 2(5) = 26$ meters

$A = b \times h = 8 \times 5 = 40$ square meters or m^2

$d = \sqrt{b^2 + h^2} = \sqrt{8^2 + 5^2} = \sqrt{89} \approx 9.4$ meters

EXAMPLE 2—

In a rectangle the base = 5 inches, and the height is 2 feet. Find the area.

Both units of measure must be the same. 2 ft = 24 in.

$A = b\,h = (5)\,(24) = 120$ square inches

EXAMPLE 3—

Find the area and the perimeter of the figure on the side.

$FE = 30$. Notice $AB + CD$ must add to 30. Since $CD = 18$, $AB = 12$.

$AGF = 20$. Again notice $AF = BC + DE$. $AF = 20$, $DE = 14$, so $BC = 6$.

The perimeter is $AB + BC + CD + DE + EF + AF = 100$.

To find the area extend BC to G, to divide the figure into two rectangles.

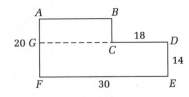

The left rectangle is $AB \times AG = 6 \times 12 = 72$.

The right rectangle is $EF \times DE = 30 \times 14 = 420$. The total is $420 + 72 = 492$ square units. Notice CG is NOT part of the perimeter.

EXAMPLE 4—

The length of a rectangle is 6 more than the width.
 If the perimeter is 28 inches, find the dimensions.

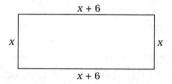

Let $x =$ width. $x + 6 =$ length. $x + x + x + 6 + x + 6 = 28$. $4x + 12 = 28$. $4x = 16$. $x = 4$, the width. $x + 6 = 4 + 6 = 10$, the length.

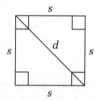

A square is a rectangle where the base $=$ height $= s$ (side).

Area $A = s \times s = s^2$. (s squared; squaring comes from a square!!!!!)

Perimeter $p = 4s$.

Diagonal $d = s\sqrt{2}$ (from the 45°–45°–90° triangle).

EXAMPLE 1—

A square is 50 miles on its side.
 Find its area, perimeter, and diagonal.

$p = 4s = 4(50) = 200$ miles

$A = s^2 = 50^2 = 2500$ square miles

$d = s\sqrt{2} = 50\sqrt{2} \approx 70.7$ miles

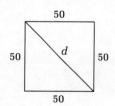

EXAMPLE 2

Find the area remaining if a square of side 9 has a square of side 7 removed.

Area remaining = area outside − area inside

$9^2 - 7^2 = 81 - 49 = 32 \text{ mi}^2$

EXAMPLE 3

The area of a square is 81. Find its perimeter.

$s^2 = 81 \qquad s = \sqrt{81} = 9 \qquad p = 4s = 4(9) = 36$

Area of a triangle. Before we talk about the area of a triangle, we must talk about its *height* or *altitude*.

The height is a line drawn from a vertex, perpendicular (right angle) to the opposite base, extended if necessary.

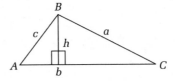

In the second picture we had to extend the base, but the extension is not part of the base.

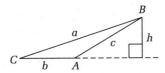

Area of a triangle $A = \dfrac{1}{2}b \times h$ (one-half base times the height, drawn to that base, extended if necessary).

Perimeter: add up the three sides, $a + b + c$. (Not h, unless it is a side!)

EXAMPLE 1—

If everything is in meters, find the area and perimeter.

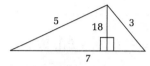

$p = 3 + 5 + 7 = 15$ meters
$A = (.5)(1.8)(7) = 6.3$ square meters

OK, OK, why the half? If you thought half a rectangle, you would be correct.

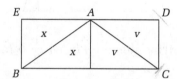

Draw the triangle and draw the rectangle around it. The x's and v's show the triangle ABC is really half rectangle BCDE. That's all there is to that.

EXAMPLE 2—

Find x.

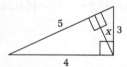

In the right triangle $A = (½)$ base × height = ½ the product of its legs. (Either could be considered base and the other the height.)

$A = (½) (3)(4) = 6$. But here is another base and height here.

$A = (½) (5)x = (5⁄2)x$ (5 is the hypotenuse and x is the height to that hypotenuse).

$(5⁄2)x = 6$. $5x = 12$. $x = 12⁄5 = 2.4$.

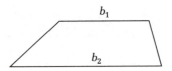

Trapezoid area = $A = \dfrac{1}{2} h (b_1 + b_2)$. Why?

Drawing one diagonal, we see a trapezoid is the sum of two triangles.

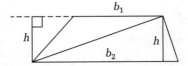

The area of the triangles is $\dfrac{1}{2} b_1 h + \dfrac{1}{2} b_2 h$.

Factoring out $(1/2)h$, we get $A = \dfrac{1}{2} h(b_1 + b_2)$.

EXAMPLE 1—

If $h = 10$ yards and the bases are 12 and 14 yards, find the area.

$A = \dfrac{1}{2} h(b_1 + b_2) = \dfrac{1}{2} (10)(12 + 14) = (5)(26)$

$= 130$ square yards.

EXAMPLE 2—

If the area is 100, the height is 4, and one base is 3, find the other base.

$A = \dfrac{1}{2} h(b_1 + b_2)$ $100 = \dfrac{1}{2} (4)(3 + x)$ $2(3 + x) = 100$

$2x + 6 = 100$ $2x = 94$ $x = 47$

EXAMPLE 3—

In an isosceles trapezoid, the upper base is 16, the lower base is 24. The height is 3. Find the perimeter (a little tough).

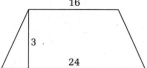

Draw the picture. $y = 16$. Since it is an isosceles trapezoid, we get $2x + 16 = 24$. $2x = 8$. $x = 4$. The slanted side is the hypotenuse of a 3, 4, . . . 5 (Pythagorean triple right triangle).

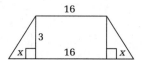

Perimeter p = top base + bottom base + two equal sides = $24 + 16 + 2(5) = 50$. Notice, the height is *not* part of the perimeter.

We'll get back to more of these later. Let's talk about circles.

CHAPTER II

ALL ABOUT CIRCLES

When you take geometry, you will spend lots of times with circles. Here we will just give some basic facts and do a few problems.

Definition *Circle:* The set of all points in a plane at a given distance from a given point.

The given point is called the *center* of the circle.

The letter that many times indicates the center of the circle is *O.*

The length of the outside of the circle, the perimeter of the circle, is called the *circumference.*

A *radius* is the distance from the center to any point on the circumference.

OA and *OB* are *radii* (plural of radius).

A *diameter* is a line segment drawn from one side of the circle through the center to the other side of the circle. *AB* is a diameter.

The diameter is twice the radius. $d = 2r$ or $r = d/2$.

π, the Greek letter pi, is the ratio of the circumference to the diameter.

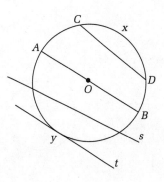

$$\pi = \frac{c}{d} \quad \text{or} \quad c = \pi d = 2\pi r \quad \pi = 3.14159265\ldots$$

or ≈ 3.14

Some people have a tough time understanding π, as opposed to cherry, apple, or peach pi. If you do, draw a circle. Take a string and cut it the length of the diameter. Lay the string, a nonelastic string, on the circumference of the circle. It goes on the circumference a little more than 3 times. So the value of pi is approximately 3.14, and 3.14 times the diameter is the circumference of the circle!!!!

The area of a circle, $A = \pi r^2$. We cannot show that this formula is true now. It requires calculus. As a matter of fact, many students are thrilled when they are finally shown what the area of a circle is and they haven't been lied to all these years.

Let's get to some other definitions.

Definition   *Chord:* A line segment drawn from one side of a circle to the other. *CD* is a chord. The diameter is the largest chord.

$\overset{\frown}{CXD}$ is an *arc*. It is a *minor arc* since it is less than ½ the circle.

$\overset{\frown}{CYX}$ is a *major arc* since it is more than ½ a circle.

$\overset{\frown}{AXB}$ and $\overset{\frown}{AYB}$ are *semicircles,* ½ a circle.

Tangent: a line, line *t*, hitting a circle in one point; *Y* is the point of tangency.

Secant: a line, line *s*, hitting a circle in two points. Let's do some problems.

EXAMPLE 1—

The radius of a circle is 7 inches. Find the area and the circumference.

$A = \pi r^2 = \pi(7)^2 = 49\pi$ square inches. $A \approx 3.14 \times 49 = 153.86$ square inches.

$c = 2\pi r = 2(3.14)7 = 43.96$ inches, approx. From now on, I won't multiply out the answers since 49π is more accurate. *Be sure you practice lots of the basic problems before you go on.*

EXAMPLE 2—

If the area is 36π find the circumference.

$$A = \pi r^2 \qquad 36\pi = \pi r^2 \qquad \frac{36\pi}{\pi} = \frac{\pi r^2}{\pi} \qquad r^2 = 36$$

$$r = \sqrt{36} = 6 \qquad c = 2\pi r = 2\pi(6) = 12\pi$$

EXAMPLE 3—

If the radius is 6 feet, find the area of the shaded figure. The part of the figure that looks like a piece of pie is called a sector. If you look at this figure, the shaded portion is the area of ¼ of the circle minus the area of the triangle.

$A = \dfrac{1}{4}\pi r^2 - \dfrac{1}{2}bh.$ The radius of the circle is 6. The base of the triangle is 6 and the height of the triangle is 6 since all radii of a circle are equal.

So $A = \dfrac{1}{4}\pi r^2 - \dfrac{1}{2}bh = \dfrac{1}{4}\pi(6)^2 - \dfrac{1}{2}(6)(6) = 9\pi - 18$ square units.

EXAMPLE 4—

Find the area and perimeter of the region if $AB = 20$ feet and $BC = 30$ feet.

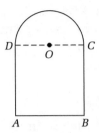

This figure is called a Norman window and is a rectangle with a semicircle (half a circle) on top. This is a big Norman window.

Draw *CD*. The total area is the area of the rectangle plus the area of half a circle. In symbols $A = bh + \dfrac{1}{2} \pi r^2$.

The base $b = AB = 20$. The height, $h = BC = 30$.
$AB = CD = \text{diameter} = 20$. So the radius is $(1/2)20 = 10$.

$$A = bh + \frac{1}{2} \pi r^2 = (20)(30) + \frac{1}{2} \pi(10)^2$$

$$= 600 + 50\pi \text{ square feet}$$

The perimeter $= AB + BC + AD + \frac{1}{2}$ the circumference of the circle. Notice *CD* is NOT part of the perimeter since it is not on the outside.

$$p = 20 + 30 + 30 + \frac{1}{2} 2\pi(10) = 80 + 10\pi \text{ feet}$$

EXAMPLE 5—

Find the area and perimeter of the sector if the angle is 60 degrees and the radius is 10 meters.

Since 60° is ⅙ of a circle (60°/360°),

$$A = \frac{1}{6} \pi r^2 = \frac{1}{6} \pi(10)^2 = \frac{50}{3} \pi \text{ square meters}$$

The perimeter consists of two radii plus an arc that is ⅙ of the circle.

$$p = 2r + \frac{1}{6}\, 2\pi r = 2(10) + \frac{1}{6}\, 2\pi(10) = 20 + \frac{10}{3}\, \pi \text{ meters}$$

As I said, in geometry we will do much more with circles. For now, let's do some 3-D stuff.

CHAPTER 12

VOLUMES AND SURFACE AREA IN 3-D

This is a topic that used to be taught a lot in fifth and sixth grades and in geometry as is needed later on. It is not difficult. You should see the picture, understand the picture, and memorize the formulas and then be able to use them.

POSTULATE The volume of a box is length × width × height or $V = lwh$.

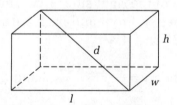

Surface area = $\quad 2\,lw \;+\; 2wh \;+\quad 2lh$

                   Bottom + top   Sides   Front + back

Diagonal (3-D Pythagorean theorem): $d = \sqrt{l^2 + w^2 + h^2}$

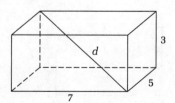

EXAMPLE I—

If the length is 7 feet, width 5 feet, height 3 feet, find the volume, surface area, and diagonal.

$V = lwh = (7)(5)(3) = 105$ cubic feet

$SA = 2lw + 2wh + 2lh = 2(7)(5) + 2(5)(3) + 2(7)(3) = 142$ square feet

$d = \sqrt{7^2 + 5^2 + 3^2} = \sqrt{49 + 25 + 9} = \sqrt{83} \approx 9.1$ feet

NOTE

The correct name for a box is rectangular parallelpiped. A parallelpiped is the 3-D figure where the opposite sides are parallelograms. Rectangular means all angles are right angles. You understand why I like to use the word "box" instead.

EXAMPLE 2 (A PEEK AT THE FUTURE)—

A box has square base, no top, and volume 100. Write in terms of one variable the surface area.

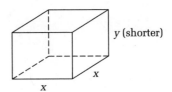

y (shorter)

Let x = length. Since it is a square base, the width is also x. Since we don't know the height and it may be different, we let y = the height.

$V = lwh = xxy = x^2y = 100$

The surface area S: no top; the bottom is a square: area is x^2. The area of the right side is xy. The left side better be the same, xy. Front and back are also xy. Total surface area $S = x^2 + 4xy$.

If $x^2y = 100$, then $y = \dfrac{x^2y}{x^2} = \dfrac{100}{x^2}$ aaand $S = x^2 + 4xy = x^2 + 4x\dfrac{100}{x^2}$. Sooo $S = x^2 + \dfrac{400}{x}$ and we have written the surface area in terms of one variable.

Like many math problems, you must see the picture.

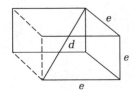

Cube: A box with the length, width, and height the same. We call them *edges.*

A cube has 12 edges, 8 *vertexes* (*vertices*), the points (where the edges meet), and 6 *faces* (sides), all squares.

Let e = edge. $V = exexe = e^3$, e cubed (cubing comes from a cube, like squaring comes from a square!!!).

Surface area: $S = 6e^2$ (6 squares) $d = s\sqrt{3}$.

EXAMPLE 1—

Find V, S, d if $e = 4$ meters.

$V = 4^3 = 64$ m^3 $S = 6e^2 = 6(4)^2 = 96$ m^2 $d = 4\sqrt{3}$ m

EXAMPLE 2—

If $S = 150$, find the volume.

$S = 150 = 6e^2$ $e^2 = \dfrac{150}{6} = 25$ $e = \sqrt{25} = 5$

$V = e^3 = 5^3 = 125$

The rest of the formulas really cannot be derived now. All require calculus.

1. If the two bases are the same, the volume is the area of the base times the height.

2. If the top comes to a point, the volume is multiplied by ⅓.

Cylinder: Base is a circle V = area of the base times the height $V = \pi r^2 h$.

There are three surfaces. The top and bottom are circles, area $2\pi r^2$. If you carefully cut the label off a can of soup, you get a rectangle. If you don't count the rim, the height of the can is the height of the rectangle. The width is the circumference of the circle!!!! Area of the curved surface (the label) is $2\pi rh$.

Total surface area: $S = 2\pi r^2 + 2\pi rh$

EXAMPLE

Find the volume and surface area of a cylinder height 4 yards, diameter of the base 20 yards.

If $d = 20$, $r = 10$. $V = \pi r^2 h = \pi(10)^2(4) = 400\pi$ yd³.
$S = 2\pi r^2 + 2\pi r h = 2\pi(10)^2 + 2\pi(10)(4) = 280\pi$ yd².

Sphere: $V = 4/3\pi r^3$. $S = 4\pi r^2$.

EXAMPLE

Find V and S if $r = 6$ centimeters.

$$V = \frac{4}{3}\pi r^3 = \frac{4}{3}\pi(6)^3 = 288\pi \text{ cm}^3 \qquad S = 4\pi r^2 = 4\pi r^2$$

$$= 4\pi(6)^2 = 144\pi \text{ cm}^2$$

Cone: $V = 1/3\pi r^2 h$. $S = \pi r^2 + \pi r l$. $l = $ lateral height $= \sqrt{r^2 + h^2}$.

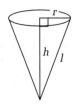

EXAMPLE

Find V and S if $h = 4$ feet and $r = 3$ feet.

$$V = \frac{1}{3}\pi r^2 h = \frac{1}{3}\pi(3)^2 4 = 12\pi \text{ ft}^3$$

$$l = \sqrt{3^2 + 4^2} = \sqrt{9 + 16} = \sqrt{25} = 5 \qquad S = \pi r^2 + \pi r l$$

$$= \pi(3)^2 + \pi(3)(5) = 24\pi \text{ ft}^2$$

Let's do two more.

> **NOTE**
>
> *I once had a neighbor who wanted to know the surface area of a can. Of course being a teacher I just couldn't tell him the answer. I had to tell him how to get it. Of course, he didn't care, but he had to listen. That is why he moved right after that. Just kidding!*

EXAMPLE I—

Find the volume and surface area of a pyramid with a square base if the side = 16, and the height is 6. Also find the length of the edge. All measurements are in feet.

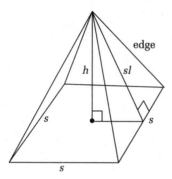

Since we have a square base and the figure comes to a point,

$$V = \frac{1}{3}\, s^2 h = \frac{1}{3}\, (16)^2(6) = 512 \text{ cubic feet}$$

The surface area consists of the base (a square) and four triangles.

$s^2 = 16^2 = 256$. Four triangles is $4((\frac{1}{2})bh)$. $b = s$. The h here is the *slant height*. The height of the pyramid is 6. If we draw a line from where the height hits the base to the side of the pyramid, that length is $(\frac{1}{2})s = 8$. The slant height is the hypotenuse of the figure = 10 (6–8–10 triple). $2sh = 2(16)10 = 320$. Total surface is $256 + 320 = 576$ square feet.

The edge of the pyramid is also the hypotenuse of a right triangle. (Try to see it.) Base = $\frac{1}{2}s = 8$. Slant height is 10. Edge = $\sqrt{8^2 + 10^2} = \sqrt{164} = 2\sqrt{41} \approx$ 12.8 feet.

EXAMPLE 2 (A LITTLE TOUGHER)—

Find V and S of the cylinder with a hemisphere (half a sphere) on top if $r = 3$ m and $h = 7$ m.

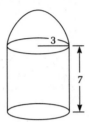

$$V = \pi r^2 h + \frac{1}{2}(\frac{4}{3}\,\pi r^3) = \pi(3)^2(7) + \frac{2}{3}\,\pi(3)^3 = 81\pi \text{ m}^3$$

There are three surfaces: the bottom, the curved cylinder side, half a sphere.

$$S = \pi r^2 + 2\pi rh + \frac{1}{2}\,4(\pi r^2) = \pi(3)^2 + 2\pi(3)(7) + 2\pi(3)^2$$

$$= 69\pi \text{ m}^2$$

Let's do a little trig.

CHAPTER 13

RIGHT ANGLE TRIGONOMETRY (HOW THE PYRAMIDS WERE BUILT)

INTRODUCTION: SIMILAR TRIANGLES

Two triangles are similar, written $\triangle ABC \sim \triangle XYZ$, if corresponding angles are equal.

Nonmathematically, two triangles are similar if they look exactly the same, but one is usually bigger or smaller.

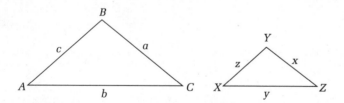

In this picture, $\angle A = \angle X$, $\angle B = \angle Y$, $\angle C = \angle Z$.

If two triangles are similar, corresponding sides are in proportion.

$$\frac{a}{x} = \frac{b}{y} = \frac{c}{z}$$

EXAMPLE—

Given that these two triangles are similar, find x and y.

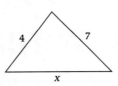 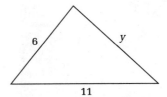

We choose the left sides first because they both have numbers. The ratio is ⅔ = ⅔. This is sometimes called the ratio of similarity.

$$\frac{\text{Small left}}{\text{Big left}} = \frac{\text{small base}}{\text{big base}} \qquad \frac{2}{3} = \frac{x}{11} \qquad 3x = 22$$

$x = 22/3$

$$\frac{\text{Small left}}{\text{Big left}} = \frac{\text{small right}}{\text{big right}} \qquad \frac{2}{3} = \frac{7}{y} \qquad 2y = 21$$

$y = 21/2$

There isn't too much more to similar triangles. Let's do trig.

TRIG

Angles are measured positively in a counterclockwise direction, because many years ago some mathematicians decided it. There is no reason. Going counterclockwise we have the first, second, third, and fourth *quadrants.* From the point (x,y) draw the right triangle, always to the x axis.

NOTE

The secret of trig is to draw triangles!

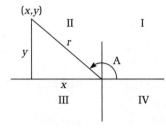

By the Pythagorean theorem, the distance from the origin $r = \sqrt{x^2 + y^2}$. r, a distance, is always positive.

Trig ratios definitions:

$$\text{sine } A = \sin A = y/r$$
$$\text{cosine } A = \cos A = x/r$$
$$\text{tangent } A = \tan A = y/x$$
$$\text{cotangent } A = \cot A = x/y$$
$$\text{secant } A = \sec A = r/x$$
$$\text{cosecant } A = \csc A = r/y$$

NOTE

These must be memorized!!

EXAMPLE 1—

Sin A = 2/9 in quadrant II. Find all the remaining trig functions.

Since sin A = 2/9, = y/r, we can let y = 2 and r = 9.

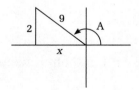

By good old Pythagoras, $x^2 + 2^2 = 81$.

$x = -\sqrt{77}$. It is minus since x is to the left. See the picture?

Now $x = -\sqrt{77}$, $y = 2$, $r = 9$. We can write out all the six trig functions!

Sin A = y/r = 2/9. Cos A = x/r = $-\sqrt{77}$ / 9.
Tan A = y/x = 2/$(-\sqrt{77})$. Cot A = x/y = $-\sqrt{77}$/2.
Sec A = r/x = 9/$(-\sqrt{77})$. Csc A = r/y = 9/2.

That's all there is. By the way because trig ratios are ratios, we can take the easiest numbers for x, y, and r. If we take larger or smaller numbers for these letters, the arithmetic would be worse, but the ratios would be the same. So we take the easiest numbers possible. In math we almost always do everything the easiest way possible.

EXAMPLE 2—

Let tan A = ¾ in III. Find sin A.

Draw the triangle up to the x axis.

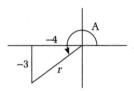

Since we are in III, both x and y are negative. Tan A = y/x = ¾.

So $x = -4$ and $y = -3$. This is a 3, 4, 5 Pythagorean triple!!!! $r = 5$!

Sin A = y/r = $-\tfrac{3}{5}$.

Notice, we could get all of them if we wanted!

EXAMPLE 3—

Csc A = −⅞. Find cot A.

Again draw the triangle to the x axis. Csc A = r/y.

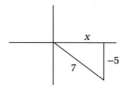

r is always + . . . so y must be −. $r = 7$ and $y = -5$.

$x = \sqrt{7^2 - (-5)^2} = \sqrt{24} = 2\sqrt{6}$ and is positive since x is to the right. Note: y positive up, y negative down. Cot A = $x/y = 2\sqrt{6}/(-5) = -2\sqrt{6}/5$.

We would like to find trig ratios of multiples of 30°–45°–60°–90°.

Again, because of trig ratios we use the easiest possible numbers.

In a 45°–45°–90° triangle, let the legs opposite each 45° angle = 1, and the hypotenuse = $\sqrt{2}$.

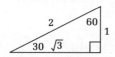

In a 30°–60°–90° triangle, let the leg opposite the 30° = 1.

The leg opposite the 60° = $\sqrt{3}$.

The hypotenuse = 2.

EXAMPLE 4—

Find the cos 120°.

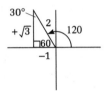

Going counterclockwise we get 90° + 30° = 120°.

The triangle to the x axis. The angle near the origin is 60° = 180° − 120°. The angle at the top is 30°.

x, opposite the 30° = −1 (minus to the left).
y, opposite the 60° = +$\sqrt{3}$ (positive, up).
r = 2 and is always positive.
Cos 120° = x/r = −½

EXAMPLE 5—

Find sec 225°.

90° + 90° = 180°   225° = 180° + 45°

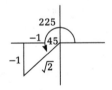

In this 45°–45°–90° triangle, both legs = −1 (both negative: x left; y down).

$r = \sqrt{2}$ and is always positive.

Sec 225° = $r/x = \sqrt{2}$ / −1 = −$\sqrt{2}$

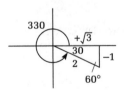

EXAMPLE 6—

Find cot 330°.

$330° = 90° + 90° + 90° + 60°$. The angle drawn is $330° = 360° - 30°$.

$y = -1, x = \sqrt{3}, r = 2 \qquad \cot 330° = x/y = \sqrt{3} / (-1) = -\sqrt{3}$

EXAMPLE 7—

Find all the trig functions of 180°.

Since this angle is on the x axis, we let $r =$ easiest number possible $r = 1$.

So this point would be $(-1,0)$. $x = -1$ and $y = 0$.

So sin 180° = $y/r = 0/1 = 0$. Cos 180° = $x/r = -1/1 = -1$. Tan 180° = $y/x = 0/1 = 0$.

Cot 180° = $x/y = 1/0$ undefined. Sec 180° = $r/x = 1/(-1) = -1$. Csc 180° = $r/y = 1/0$ undefined.

NOTE

All multiples of 90°: two trig functions = 0; two are undefined; two are either both = −1 or both = +1.

RIGHT ANGLE TRIG

We now bring all triangles to the first quadrant and restrict our discussion to sine, cosine, and tangent.

Sin A = y/r = opposite/hypotenuse.

Cos A = x/r = adjacent/hypotenuse.

Tan A = y/x = opposite/adjacent.

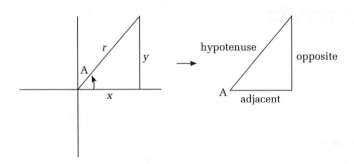

Since the angles are not necessarily the nice ones, we will need to use a calculator.

EXAMPLE 1—

Find sin 34°. In the TI 83 +, press mode key to make sure the angles are in degrees. (Later on you will use other measures of angles) sin 34° = .5591 . . .

EXAMPLE 2—

If tan A ≈ 1.2345. Find A. Press second tan. We get $\tan^{-1}$ which is the arc tan or inverse tan. (See *Precalc with Trig for the Clueless* if you want to understand more.) It gives you the angle. $A = 50.9909 . . .°$

EXAMPLE 3—

$A = 62°$. $AC = 25$. Angle C is the right angle.

Find all the remaining parts of the triangle.

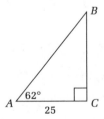

Angle $B = 90° - 62° = 28°$. Tan 62° = $BC/25$. $BC = 25$ tan $62° = 47$. $AB = \sqrt{25^2 + 47^2}$ or cos 62° = $25/AB$. $AB = 25/\cos 62°$. $AB = 53$.

EXAMPLE 4—

The angle of depression to a boat is 22°. If the light-house is 200 feet tall, find the distance from the light-house to the boat.

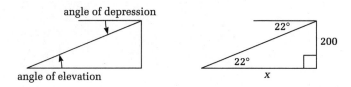

Angle of elevation = angle of depression (alternate interior angles). Tan 22° = 200/x. x = 200 / tan 22° ≈ 495 feet.

EXAMPLE 5—

Find the height of the antenna.

Tan 20° = y/80. y = 80 tan 20° ≈ 29.

Tan 52° = (z + y)/80. z + y = 80 tan 52° ≈ 102.

x = 102 − 29 = 73-foot antenna (pretty big).

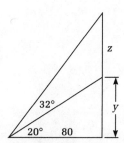

I hope you find basic trig easier now. Let's go over several other topics.

CHAPTER 14

MISCELLANEOUS

Here are a few more topics that you may encounter.

SETS

A *set* is a collection of things. Each "thing" is called an *element*.

A set is denoted by braces { }. Capital letters are used for sets. Small letters are used for elements.

$S = \{5, 7, 9, 2\}$ is a set with four elements. We write $2 \in S$ (2 is an element in the set S).

$4 \notin S$, 4 is not an element of S. $\in$ is the Greek letter epsilon.

Believe it or not, we probably could get away with just this, but let's do a little more.

EXAMPLE—

$\{3, 4\} = \{4, 3\}$.

Order does not count in a set.

$\{5, 7\} = \{7, 7, 5, 5, 7, 5, 5, 5\}$.

Repeated elements aren't counted. Each set here has two elements, 5 and 7.

U = universe, the set of everything we are talking about.

A^c read "A complement" = all elements in the universe not in A.

$\{ \}$ or ϕ (Greek letter phi, pronounced fee or fie) the set with no elements

For example the set of all 127-foot human beings.

$A \cup B$, "A union B", is the set of all elements in A or in B or in both (everything in A and B put together.

$A \cap B$, "A intersect(ion) B", the set of all elements in A and also in B

Let $U = \{1,2,3,4,5,6,7,8,9\}$ $A = \{1,2,3,4,5,\}$
$B = \{3,5,7,9\}$ $C = \{1,4\}$

EXAMPLE 1

$A \cup B = \{1,2,3,4,5,7,9\}$ Elements are put in only once (in any order) (like uniting the 2 sets).

EXAMPLE 2

$A \cap B = \{3,5\}$ elements in both A and B (like the intersection of 2 streets).

EXAMPLE 3

$B \cap C = \varnothing$ B and C are *disjoint* sets, sets with nothing in common.

EXAMPLE 4

$A^c = \{6,7,8,9\}$ Everything in the universe not in A.
Change the universe and you change the complement.

Let U = the letters in the alphabet $\{a, b, c, \ldots ,z\}$.
Vowels $V = \{a, e, i, o, u\}$. C = consonants.

NOTE

You can't say y is sometimes a vowel. In this case we will say y is NOT a vowel.

EXAMPLE 5

$V \cup C = U$.

EXAMPLE 6

$V^c = C$ and $C^c = V$.

EXAMPLE 7

V and C are disjoint.

That is really all you need about sets.

FUNCTIONS

This is truly an important topic. We will introduce it here. One of the few good things about the newer math is there is more time spent on functions (still not enough, but better).

Definition   Given a set D. To each element in D we assign one and only one element

EXAMPLE 1

$$1 \longrightarrow r$$
$$2 \longrightarrow 2$$
$$3 \nearrow$$
$$4 \longrightarrow \text{cow}$$
$$D$$

Is this a function? To each element in D have I assigned one and only one element? To the number 1, we've assigned r. To the number 2 we've assigned 2. To the number 3, we've assigned 2. To the number 4, we've assigned a cow. To each element in D, we've assigned one and only one element. This IS a function.

The set D is called the domain.

Although not part of the definition, there is always a second set, called the range

For the domain, we usually think of the x values. For the range, we think of the y values.

The rule (in this case the arrows) is called a map or mapping. 1 is mapped into r; 2 is mapped into 2; 3 is mapped into 2; 4 is mapped into a cow.

NOTE

Two numbers in D can be assigned the same element

ALSO NOTE

There can be elements in common between the domain and range (the number 2) or they can be totally different (like a number and a cow).

EXAMPLE 2

Is this a function?

$$a \longrightarrow b$$
$$b \longrightarrow c$$

No, this is not a function since a is assigned to two values, b and c.

This notation is kind of messy. There are a number of possible notations. We will use the most common.

EXAMPLE 3

(Example 1, revisited) $f(1) = r$ (f of 1 is or equals r). $f(2) = 2$. $f(3) = 2$. $f(4) = $ cow. IMPORTANT! $f(1)$ does not mean multiplication.

NOTE

Another notation $f{:}1 \to r$ (f takes 1 into r); $f{:}2 \to 2$; $f{:}3 \to 2$ $f{:}4 \to$ cow.

Normally, the map is listed in formula form.

EXAMPLE 4—

$f(x) = x^2 + 3x + 1$. Domain $D = \{5, 2, -4\}$.

$f(5) = 5^2 + 3(5) + 1 = 41$ $f(2) = 2^2 + 3(2) + 1 = 11$
$f(-4) = (-4)^2 + 3(-4) + 1 = 5$

EXAMPLE 5—

Graph $f(x) = x^2$. Make a table.

x	$f(x)$	$(x, f(x))$
-3	9	$(-3, 9)$
-2	4	$(-2, 4)$
-1	1	$(-1, 1)$
0	0	$(0, 0)$
1	1	$(1, 1)$
2	4	$(2, 4)$
3	9	$(3, 9)$

> **NOTE**
>
> If the domain is not given, it is as large as it possibly can be.

The graph looks like this:

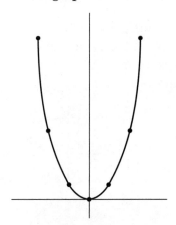

LOTS OF NOTES

1. The graph is a parabola. We will deal a lot with parabolas as we go on in math, a little more later in the book. V is the *vertex.*

2. Instead of graphing the points (x,y), we are graphing $(x, f(x))$: same idea, different notation.

3. In *Algebra for the Clueless* and beyond, we show you what values for x to choose.

EXAMPLE 6—

Graph $f(x) = \sqrt{x - 3}$. Here, we choose numbers to give us exact square roots.

x	$f(x)$	$(x, f(x))$
3	0	(3,0)
4	1	(4,1)
7	2	(7,2)
12	3	(12,3)

The graph looks like this:

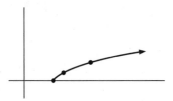

You can tell a function by the vertical line test. If all vertical lines hit the curve in one and only one place, this means for each x value there is one and only one y value. Otherwise it is not a function.

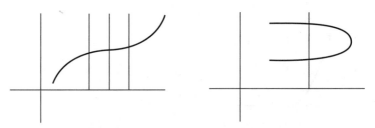

A function   Not a function
All lines are functions except vertical lines.

EXAMPLE 5—

This parabola is a function, but the one on the side is not.

yes   no

This is only a small beginning of a very important topic. At every stage of algebra, you will get more.

LINEAR INEQUALITIES

This topic is a nice topic, and necessary, but the first time you really need it is when you talk about functions.

$a < b$ (a is less than b) if a is to the left of b on the number line.

$b > a$ (b is greater than a) if b is to the right of a on the number line.

$a < b$ is the same as saying $b > a$.

$a \le b$: a is less than or equal to b.

Is $6 \le 8$? Yes, $6 < 8$.

Is $2 \le 2$? Yes, $2 = 2$. $\le$ is true if we have $<$ or $=$!!

Is $6 \le 4$? No $6 > 4$ (or $6 \ge 4$).

The following are the rules for inequality.

1. Trichotomy: Exactly one is true: $a > b$, $a = b$, $a < b$. Translation: in comparing two numbers, the first is either greater than, equal to, or less than the second.

2. Transitivity: If $a < b$ and $b < c$, then $a < c$. Translation: If the first is less than the second and the

second is less than the third, the first is less than the third. Also, transitivity holds for $\leq$, $>$, $\geq$, but I'm not writing this four times.

3. If $a < b$ (also $\leq$, $>$, $\geq$), the following are true:

Letters	One Numerical Example		
	$a < b$	$6 < 9$	
	$a + c < b + c$	$6 + 2 < 9 + 2$	
	$a - c < b - c$	$6 - 4 < 9 - 4$	
$c > 0$	$ac < bc$	$6(2) < 9(2)$	$c > 0$: c is positive
$c > 0$	$a/c < b/c$	$6/3 < 9/3$	
$c < 0$	$ac > bc$	$6(-2) > 9(-2)$ -12 is to the right of -18	$c < 0$ c is negative
$c < 0$	$a/c > b/c$	$6/(-3) > 9/(-3)$ -2 is bigger than -3	

Translation   You solve inequalities in exactly the same way you solve equations, except when you multiply or divide by a negative, you change the direction of the inequality.

EXAMPLE 1

Solve for x and graph

$$4x + 6 < 2x - 8$$

$$ -6 \quad -6$$

$$4x < 2x - 14 \text{ (order doesn't change)}$$

$$-2x \quad -2x$$

$$2x < -14 \text{ (order doesn't change)}$$

$$(2x)/2 < (-14)/2$$

$$x < -7 \text{ (order doesn't change since we divided by a positive)}$$

Graph

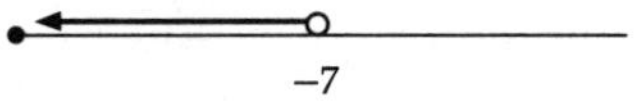

The graph as an open dot (○) on –7 is not part of the answer.

EXAMPLE 2—

$$
\begin{aligned}
4(2x-5)-3(4x-2) &\le 10\\
8x-20-12x+6 &\le 10\\
-4x-14 &\le 10\\
+14 &\quad\; 14\\
-4x &\le 24\\
(-4x)/(-4) &\ge (24)/(-4) \quad \text{(order switches since we divided by a negative)}\\
x &\ge -6
\end{aligned}
$$

Graph

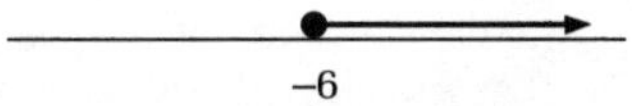

There is a solid dot (●) on –6 since –6 IS part of the answer.

Let's show when you might use inequalities for the first time.

EXAMPLE 3—

Find the domain of $f(x) = \sqrt{x-8}$.

We know you cannot have the square root of a negative number. So

$x - 8 \ge 0$

Adding 8 to each side we get $x \ge 8$, which is the domain of this $f(x)$.

ABSOLUTE VALUE

Absolute value is a topic I like to leave to this point since it seems very easy but is not. There are two definitions. We need to give both.

Definition 1   Absolute value: $|x| = \sqrt{x^2}$.

$|6| = \sqrt{(6)^2} = \sqrt{36} = 6$ $\qquad$ $|-4| = \sqrt{(-4)^2} = \sqrt{16} = 4$
$|0| = \sqrt{0^2} = \sqrt{0} = 0$.

Although this is the easier one, the other is the more usable.

Definition 2

$$|x| = \begin{array}{l} x \text{ if } x > 0 \\ -x \text{ if } x < 0 \\ 0 \text{ if } x = 0 \end{array}$$

$|6| = 6$ since $6 > 0$. $|-4| = -(-4) = 4$ since $-4 < 0$.
Read definition 2 and the example very carefully!

NOTE

$|x|$ is never negative.

EXAMPLE 1—

Solve for x: $|2x - 5| = 7$.

If $|u| = 7$, then $u = \pm 7$ $\qquad$ $|2x - 5| = 7$
means $\qquad$ $2x - 5 = 7$ or $2x - 5 = -7$

Solving we get $x = 6$ and $x = -1$.

Check

$|2(6) - 5| = |7| = 7$. $|2(-1) - 5| = |-7| = 7$

EXAMPLE 2—

$|2x - 5| = 0$. $2x - 5 = 0$. $x = 5/2$.

Check:

$|2(5 / 2) - 5| = |0| = 0$

EXAMPLE 3—

$|3x + 7| = -2$. There is no answer since the absolute value is never negative.

EXAMPLE 4—

Solve for x and graph: $|x - 6| \leq 4$.

We are looking for all integers within 4 of 6. If u were an integer and $|u| \leq 4$, the u could be 4, 3, 2, 1, 0, −1, −2, −3, −4. In other words $-4 \leq u \leq 4$ (u is between −4 and 4 inclusive).

$|x - 6| \leq 4$ means $-4 \leq x - 6 \leq 4$

$+6 \qquad +6 \ +6$ (add 6 to all three parts)

$2 \leq x \leq 10$

Notice   All of these numbers are within 4 of 6.

EXAMPLE 5—

Solve for x and graph:

$|2x - 5| < 11$

$-11 < 2x - 5 < 11$

$+5 \qquad +5 \ +5$

$-6 < 2x < 16$

$-3 < x < 8$

EXAMPLE 6—

Solve for x and graph: $|x - 8| \geq 3$

We are looking for all numbers 3 or more away from 8. Such numbers would be 3, 4, 5, 6, . . . ($x - 8 \geq 3$) or

−3, −4, −5, −6, . . . ($x − 8 ≤ −3$) since the absolute value of all these numbers is greater than or equal to 3. To summarize, $|x − 8| ≥ 3$ means $x − 8 ≥ 3$ or $x − 8 ≤ −3$.

We get $x ≥ 11$ or $x ≤ 5$. The graph is:

Notice again the answers are all 3 or more away from 8.

EXAMPLE 7—

Solve for x and graph: $|5 − 3x| ≥ 7$.

$|5 − 3x| ≥ 7$ means $5 − 3x ≥ 7$ or $5 − 3x ≤ −7$

$−5$ $−5$ $−5$ $−5$

$−3x ≥ 2$ $−3x ≤ −12$

$x ≤ −2 / 3$ or $x ≥ 4$

(Divide by negatives? The direction of inequality switches.)

EXAMPLE 8—

$|5x + 1| < −4$. No answers since the absolute value is never negative.

EXAMPLE 9—

$|−3x − 8| ≥ −2$. The answer is all real numbers since the absolute value is always bigger than any negative number (since the absolute value is 0 or positive).

Let's take a look at matrices.

MATRICES

Matrices (plural of matrix) are arrays (in this case, of numbers or letters).

m × n matrix read, "**m** by **n** matrix" has **m** rows and **n** columns.

EXAMPLE 1

$\begin{bmatrix} a & b & c & d \\ e & f & g & h \end{bmatrix}$ is a 2 (rows) by 4 (columns) matrix:

2 rows: [$a\ b\ c\ d$] and [$e\ f\ g\ h$] and

4 columns: $\begin{bmatrix} a \\ e \end{bmatrix}, \begin{bmatrix} b \\ f \end{bmatrix}, \begin{bmatrix} c \\ g \end{bmatrix}, \begin{bmatrix} d \\ h \end{bmatrix}$.

a is the 1st row 1st column entry. e is the 2d row 1st column entry.

b is the 1st row 2d column entry. f is the 2d row 2d column entry.

c is the 1st row 3d column entry. g is the 2d row 3d column entry.

d is the 1st row 4th column entry. h is the 2d row 4th column entry.

To add, both must have the same number of rows and columns. Add the same entry.[]

EXAMPLE 2

$$\begin{bmatrix} a & b \\ c & d \\ e & f \end{bmatrix} + \begin{bmatrix} g & h \\ i & j \\ k & l \end{bmatrix} = \begin{bmatrix} a+g & b+h \\ c+i & d+j \\ e+k & f+l \end{bmatrix}$$

EXAMPLE 3

$$\begin{bmatrix} 5 & 2 & -3 & 0 & 1 \\ -2 & 4 & 5 & 1 & 6 \end{bmatrix} + \begin{bmatrix} 2 & 3 & 4 & 5 & 6 \\ 1 & 3 & -1 & -2 & -3 \end{bmatrix} = \begin{bmatrix} 7 & 5 & 1 & 5 & 7 \\ -1 & 7 & 4 & -1 & 3 \end{bmatrix}$$

In multiplication, if we multiply an **m** $\times$ **n** matrix by an **n** $\times$ **p** matrix, we get an **m** $\times$ **p**. In other words, if the columns of the first matrix are the same as the rows of the second matrix, we can multiply. We get a matrix which is the row of the first matrix by the column of the second matrix.

EXAMPLE 4

$$\begin{bmatrix} a\ b\ c\ d \\ e\ f\ g\ h \end{bmatrix} \times \begin{bmatrix} o\ p\ q \\ r\ s\ t \\ u\ v\ w \\ x\ y\ z \end{bmatrix}$$ 1st 2×4; 2d 4×3. We can multiply

them. The answer will be 2×3.

1st row 1st column entry: The sum of the 1st row in 1st matrix multiplied by the 1st column in 2d matrix: $ao + br + cu + dx$.

1st row 2d column entry: The sum of the 1st row in 1st by 2d column in 2d: $ap + bs + cv + dy$.

1st row 3d column entry: The sum of the 1st row in 1st by 3d column in 2d: $aq + bt + cw + dz$.

2d row 1st column entry: The sum of the 2d row by 1st column: $eo + fr + gu + hx$.

2d row 2d column entry: The sum of the 2d row by 2d column: $ep + fs + gv + hy$.

2d row 3d column entry: The sum of the 2d row by 3d column: $eq + ft + gw + hz$.

The matrix would look like this:

$$\begin{bmatrix} ao + br + cu + dx & ap + bs + cv + dy & aq + bt + cw + dz \\ eo + fr + gu + hx & ep + fs + gv + hy & eq + ft + gw + hz \end{bmatrix}$$

You cannot multiply them in the reverse order.

NOTE

I put so many letter problems here because with letters, in this case, you can see what has happened better than with numbers.

EXAMPLE 5

Let $\mathbf{A} = \begin{bmatrix} 1\ 2 \\ 3\ 4 \end{bmatrix}$ and $\mathbf{B} = \begin{bmatrix} 5\ 6 \\ 7\ 8 \end{bmatrix}$. Find $\mathbf{AB}$ and $\mathbf{BA}$.

Both are 2 by 2; so we can multiply both ways.

$$\begin{bmatrix} 1\ 2 \\ 3\ 4 \end{bmatrix}\begin{bmatrix} 5\ 6 \\ 7\ 8 \end{bmatrix} = \begin{bmatrix} 1(5) + 2(7) & 1(6) + 2(8) \\ 3(5) + 4(7) & 3(6) + 4(8) \end{bmatrix} = \begin{bmatrix} 19\ 22 \\ 43\ 50 \end{bmatrix} = \mathbf{AB}$$

$$\begin{bmatrix} 5 & 6 \\ 7 & 8 \end{bmatrix}\begin{bmatrix} 1 & 2 \\ 3 & 4 \end{bmatrix} = \begin{bmatrix} 5(1)+6(3) & 5(2)+6(4) \\ 7(1)+8(3) & 7(2)+8(4) \end{bmatrix} = \begin{bmatrix} 23 & 34 \\ 31 & 48 \end{bmatrix} = \mathbf{BA}$$

Also note $\mathbf{AB} \neq \mathbf{BA}$. In other words multiplication of matrices is NOT commutative. This leads us naturally into the next section.

FIELD AXIOMS AND WRITING THE REASONS FOR THE STEPS TO SOLVE EQUATIONS

Axiom: a law taken to be true without proof.

This is a very important section to go over "very gently." You should know the laws in the chapter, but going over this section in too much depth at this point has led more than one student to dislike math. The one "problem" we will do at the end of this section is the reason for this dislike.

Field

Given a set S, we have two operations, $+$ and $\times$. $a, b, c \in S$.

1.,2. The closure laws: If $a, b \in S$, $a + b \in S$ and $a \times b \in S$.

Translation: If we take any two elements in the set, even the same one twice, and if we add them together, the answer is always in the set, and when we multiply them together, the answer is always in the set. We need one example.

Let $S = \{0, 1\}$ The set is closed under multiplication since $0 \times 0 = 0$, $1 \times 0 = 0$, $0 \times 1 = 0$, $1 \times 1 = 1$: the answer is always in the set. The set is not closed (the opposite of closed) under multiplication since $0 + 0 = 0$, $1 + 0 = 1$, $0 + 1 = 1$, but $1 + 1 = 2$ and $2 \notin S$.

3.,4. The commutative laws of addition and multiplication:

$a + b = b + a$ and $a \times b = b \times a$ for all $a, b \in S$
$3 + 5 = 5 + 3$ and $9(8) = 8(9)$

5.,6. The associative laws of addition and multiplication:

$$(a + b) + c = a + (b + c) \qquad (a \times b) \times c = a \times (b \times c)$$
for all $a, b, c \in S$

7. Identity for addition: There is a number, call it 0, such that $a + 0 = 0 + a = a$ for all $a \in S$.

8. Identity for multiplication: There is a number, called 1, such that $1 \times a = a \times 1 = a$ for all $a \in S$.

9. Inverse for +: For all $a \in S$, there is a number $-a$, such that $a + (-a) = (-a) + a = 0$.

$5 + (-5) = (-5) + 5 = 0$

10. Inverse for x: For all $a \neq 0 \in S$, there is a number $1/a$ such that $(1/a)(a) = a(1/a) = 1$.

11. Distributive law: For all $a, b, c \in S$, $a \times (b + c) = a \times b + a \times c$.

$5(2x - 7) = 10x - 35$

We need to know four more laws: If $a = b$, then:

1. $a + c = b + c$: If equals are added to equals, their sums are equal.

2. $a - c = b - c$: If equals are subtracted from equals, their differences are equal.

3. $a \times c = b \times c$: If equals are multiplied by equals, their products are equal.

4. $a/c = b/c$ $c \neq 0$: If equals are divided by nonzero equals, their quotients are equal.

We might need four more later.

NOTE 1

A field must have all 11 properties.

NOTE 2

The real numbers: any number that can be written as a decimal forms a field.

NOTE 3

Although it is nice to know all of them, the most important to this point is the distributive law.

1. Reflexive law: $a = a$: A quantity always equals itself.

2. Symmetric law: If $a = b$, then $b = a$. In an equation, if the left side equals the right, then we can switch sides. For example, if $2x + 3 = 5x + 7$, then $5x + 7 = 2x + 3$.

3. Transitive law: If $a = b$ and $b = c$, then $a = c$. (There are lots of operations that are transitive.)

4. Substitution: If $a = b$, then a may be substituted for b in any mathematical statement.

> **NOTE 4**
>
> *The commutative and associative laws say you can add or multiply in any order and you get the same answer.*

Whew!!! The bottom eight you will use a lot in geometry. The first 11 you will use mostly informally (except the distributive law) in algebra.

Here is a sample problem. Solve for x and explain all the steps: $5x - 4 = 2x + 17$

Statement	Reason
1. $5x - 4 = 2x + 15$	1. Given.
2. $(5x - 4) - 2x = (2x + 15) - 2x$	2. If equals are subtracted from equals, their differences are equal.
3. $-2x + (5x - 4) = -2x + (2x + 15)$	3. Commutative law of + on both sides.
4. $(-2x + 5x) - 4 = (-2x + 2x) + 15$	4. Associative law of addition on both sides.
5. $3x - 4 = 0 + 15$	5. Algebraic fact on left; additive inverse on the right.
6. $3x - 4 = 15$	6. Additive identity on the right.
7. $(3x - 4) + 4 = 15 + 4$	7. If equals are added to equals, their sums are equal.
8. $3x + (-4 + 4) = 19$	8. Associative + on the left; math fact on the right.

9. $3x + 0 = 19$	9. Additive inverse.
10. $3x = 19$	10. Additive identity.
11. $(\frac{1}{3})(3x) = (\frac{1}{3})19$	11. If equals are multiplied by equals, their products are equal.
12. $((\frac{1}{3})3)x = 19/3$	12. Associative law x on left; math fact on the right.
13. $1x = 19/3$	13. Multiplicative inverse.
14. $x = 19/3$	14. Multiplicative identity.

To understand this problem, you should read it slowly and carefully to understand each step before you go on.

TRANSFORMATIONS: PART 1

This first part is not too bad. It's not the way I like to approach things, but it is OK.

EXAMPLE 1—

We already know that $y = x^2$ looks like this:

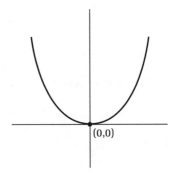

(0,0)

We sketched it by taking $x = -3, -2, -1, 0, 1, 2, 3$.

EXAMPLE 2

Sketch $y = x^2 + 4$.

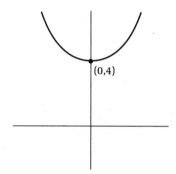

It is 4 up (almost good enough to be a soft drink).
You can see it by again letting $x = -3, -2, -1, 0, 1, 2, 3$.

EXAMPLE 3

Sketch $y = x^2 - 2$.

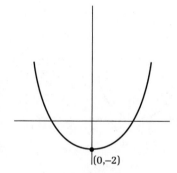

This is Example 1 but 2 steps down. Again you can
see it by taking the same x values.

EXAMPLE 4—

Sketch $y = (x - 3)^2$.

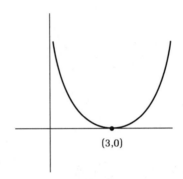

(3,0)

In sketching this it is 3 steps to the right. Take $x = 0$, 1, 2, 3, 4, 5, 6.

EXAMPLE 5—

Sketch $y = (x + 6)^2$.

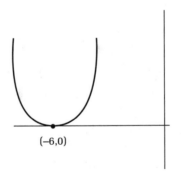

(−6,0)

This one is 6 spaces to the left. Take $x = -9, -8, -7,$ $-6, -5, -4, -3$.

EXAMPLE 6—

Sketch $y = -x^2$.

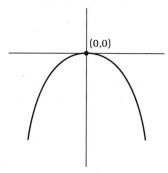

This is Example 1 upside down. Take the same values for x as in Example 1.

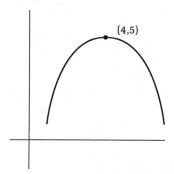

EXAMPLE 7—

A combination: Sketch $y = -(x - 4)^2 + 5$: four spaces to the right $(x - 4)$; five spaces up, +5; upside down (the minus sign). You could take the points $x = 1, 2, 3, 4, 5, 6, 7$.

EXAMPLE 8—

Compare $y = x^2$, $y = 1/2x^2$, $y = 2x^2$.

Let's actually plug in points:

x	$1/2x^2$	x^2	$2x^2$
−3	9/2	9	18
−2	2	4	8
−1	1/2	1	2
0	0	0	0
1	1/2	1	2
2	2	4	8
3	9/2	9	18

The sketches look like this:

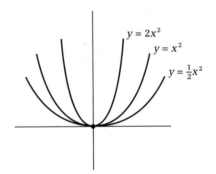

TRANSFORMATIONS: PART 2

Translations, Stretches, Contractions, Flips

Suppose $f(x)$ looks like this:

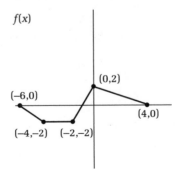

EXAMPLE 1—

f(*x*) + 5 would be 5 units up in the *y* direction. It looks like this:

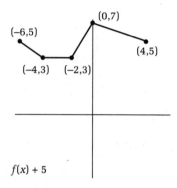

EXAMPLE 2—

f(*x*) − 3, is the original graph. Three units down it looks like this:

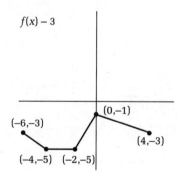

EXAMPLE 3—

$2f(x)$: y values twice as high or twice as low (points on the x-axis stay the same).

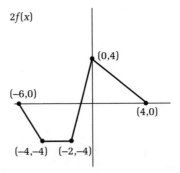

EXAMPLE 4—

$(\frac{1}{2})f(x)$: y values half as high or half as low (points on the x-axis stay the same).

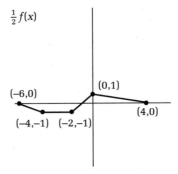

EXAMPLE 5—

$-f(x)$: upside down. Flip on the x-axis: y values that were above are below the x-axis and those below are above (the same distance from the x-axis).

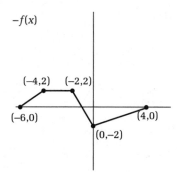

EXAMPLE 6—

$f(x + 4)$: Four units to the left.

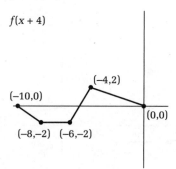

EXAMPLE 7—

$f(x - 2)$: Two units to the right.

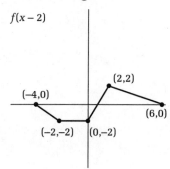

EXAMPLE 8—

$f(-x)$: flip on the y-axis. Right becomes the left and the left becomes the right (the same distance from the y-axis); y values stay the same.

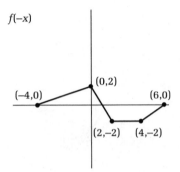

EXAMPLE 9—

$f(\frac{1}{2}x)$: Stretch twice as long: Points on y-axis stay the same.

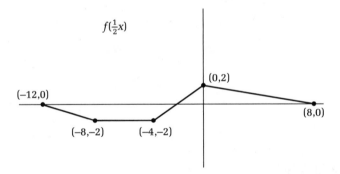

EXAMPLE 10—

$f(2x)$: compression; twice as close: points on y-axis again stay the same.

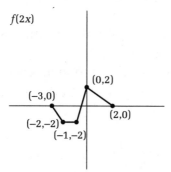

There are combos: $y = -2f(x - 1) + 3$.

1. Inside parenthesis, one unit to the right.

2. $-2f(x - 1)$, twice as high or low, upside down.

3. +3, up 3.

1.

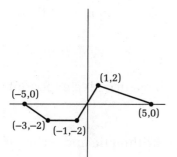

2.

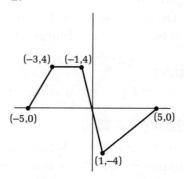

3.

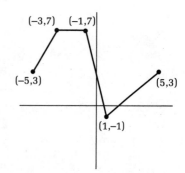

SCIENTIFIC NOTATION

Despite what I like, you probably will use your calculator. So you need to know scientific notation.

Scientific Notation   A number written in the form $b \times 10^n$, $1 \leq b < 10$, n is an integer.

   We need to do negative exponents, a topic you will do a little later.

Definition

$$x^{-n} = \frac{1}{x^n} \qquad 10^{-3} = \frac{1}{10^3}.$$

EXAMPLE 1—

Write 43,200 in scientific notation: $= 4.32 \times 10^4$. Why? $4.32 \times 10^4 = 4.32 \times 10,000 = 43,200$.

EXAMPLE 2—

Write .000000000076 in scientific notation. Eleven places to the right 7.6×10^{-11}.

   On the calculator it would come up 7.6 E – 11. E for exponent the base 10 is not included.

NOTE

Numbers that are less than one have negative exponents.

EXAMPLE 3—

Using scientific notation, do all arithmetic and write your answer in scientific notation:

$$\frac{8,000,000,000 \times .008}{20,000 \times .02} = \frac{8 \times 10^9 \times 8 \times 10^{-3}}{2 \times 10^4 \times 2 \times 10^{-2}} = \frac{\overset{4}{\cancel{8}} \times \overset{4}{\cancel{8}}}{\underset{1}{\cancel{2}} \times \underset{1}{\cancel{2}}} \times$$

$$\frac{10^9 \times 10^{-3}}{10^4 \times 10^{-2}} \text{ (Note: } 10^9 \times 10^{-3} = 10^{9-3} = 10^6 \text{)}$$

$$= 16 \times \frac{10^6}{10^2} = 16 \times 10^4.$$ But, 16 is not scientific notation!

So $16 \times 10^4 = 1.6 \times 10^1 \times 10^4 = 1.6 \times 10^5$.

That's about all for now. My hope is that you are now ready for algebra and *Algebra for the Clueless*. Good-bye for now!

INDEX

ABOUT BOB MILLER . . . IN HIS OWN WORDS

I received my B.S. and M.S. in math from Polytechnic University, Brooklyn, New York, after graduating from George W. Hewlett High School, Hewlett, Long Island, New York. After teaching my first class, substituting for a full professor, one student told another upon leaving that "at least now we have someone who can teach the stuff." I was forever hooked on teaching. Since Poly, I have taught at Westfield State, Rutgers, and I am starting my thirty-fourth year at the City College of New York. No matter how I feel, I feel better when I teach. I am always delighted when students tell me they hated math before and could never learn it, but taking math with me has made it understandable and even a little enjoyable. I will be included in the next edition of *Who's Who Among America's Teachers*. I have a fantastic wife, Marlene; a wonderful daughter, Sheryl; a terrific son, Eric; a great son-in-law, Glenn; and a great daughter-in-law, Wanda. The newest members of my family are my adorable brilliant granddaughter, Kira Lynn, 4½ years old; a brilliant, handsome grandson, Evan Ross, almost 2½; and the newest member, grandson Sean Harris, 8 hours,

1 minute old at this writing. My hobbies are golf, bowl-ing, bridge, and crossword puzzles. My practical dream is to have someone sponsor a math series from prealgebra through calculus so everyone in our coun-try will be able to think well and keep our country number one forever.

To me, teaching math is always a great joy. I hope I can give you some of this joy.